Collins

Edexcel GCSE 9-1
Maths Higher
Workbook

Trevor Senior

Preparing for the GCSE Exam

Revision That Really Works

Experts have found that there are two techniques that help you to retain and recall information and consistently produce better results in exams compared to other revision techniques.

It really isn't rocket science either – you simply need to:

- **test yourself** on each topic as many times as possible
- **leave a gap** between the test sessions.

Three Essential Revision Tips

1. Use Your Time Wisely

- Allow yourself plenty of time.
- Try to start revising at least six months before your exams – it's more effective and less stressful.
- Don't waste time re-reading the same information over and over again – it's not effective!

2. Make a Plan

- Identify all the topics you need to revise.
- Plan at least five sessions for each topic.
- One hour should be ample time to test yourself on the key ideas for a topic.
- Spread out the practice sessions for each topic – the optimum time to leave between each session is about one month but, if this isn't possible, just make the gaps as big as you can.

3. Test Yourself

- Methods for testing yourself include: quizzes, practice questions, flashcards, past papers, explaining a topic to someone else, etc.
- Don't worry if you get an answer wrong – provided you check what the correct answer is, you are more likely to get the same or similar questions right in future!

Visit **collins.co.uk/collinsGCSErevision** for more information about the benefits of these techniques, and for further guidance on how to plan ahead and make them work for you.

Command Words used in Exam Questions

This table defines some of the most commonly used command words in GCSE exam questions.

Command word	Meaning
Write... Write down...	No working will be required
Find...	A small amount of working will be required
Work out...	Some working will be required
Calculate...	Some working will be required and it is likely a calculator will be needed
Expand...	Remove brackets
Explain...	An explanation (a sentence or mathematical statement) is required
Factorise fully...	Suggests that there is more than one factor to consider (i.e. there are at least two stages in the factorisation)
Show...	All working leading to the answer must be shown
Prove...	All steps must be shown and, for a geometrical proof, reasons must be given
Justify...	Show all working or give a written explanation
Simplify fully...	Suggests more than one stage is needed to simplify an expression
Draw...	Suggests accuracy is important
Sketch...	No accurate measurements are needed

Contents

 Number   Algebra  Geometry and Measures

 Statistics  Probability Ratio, Proportion and Rates of Change

Contents 3

Order and Value

Grade 4–6 **1** Given that 5.3 × 470 = 2491, work out the value of:

a) 0.53 × 0.47

_____ [1]

b) 24.91 ÷ 0.47

_____ [1]

Grade 4–6 **2** $(15)^3 = 3375$

Find the value of $(0.15)^3$ correct to 2 significant figures.

_____ [1]

Grade 4–6 **3** Work out 27.4 ÷ 0.5

_____ [1]

Grade 4–6 **4** **a)** Write the number 35 million in standard form. _____ [1]

b) Write $3.49 × 10^{-3}$ as an ordinary number. _____ [1]

Grade 4–6 **5** These numbers are written in standard form.

$1.7 × 10^4$ $\qquad\qquad$ $4.2 × 10^3$ $\qquad\qquad$ $3.1 × 10^5$ $\qquad\qquad$ $6.5 × 10^2$

a) Work out the smallest answer that can be made from adding two of the numbers together.

_____ [2]

b) Work out the largest answer that can be made from squaring one of the numbers.

_____ [2]

Grade 4–6 **6** Work out $(4 × 10^8) × (7.5 × 10^{-2})$. Give your answer in standard form.

_____ [2]

Grade 4–6 **7** Work out $(2 × 10^6) ÷ (4 × 10^2)$. Give your answer in standard form.

_____ [2]

Total Marks _____ / 14

Types of Number

Grade 4–6 **1** $A = 3^4 \times 5$ $B = 2 \times 3^3 \times 5^2$

Write down the highest common factor (HCF) of A and B.

_____ [1]

Grade 4–6 **2**   Find the lowest common multiple (LCM) of 60 and 72.

_____ [1]

Grade 4–6 **3**   Find the highest common factor (HCF) of 84 and 120.

_____ [3]

Grade 4–6 **4**   Write 36 as a product of prime factors.

_____ [2]

Grade 4–6 **5**   Work out the value of $5^3 - 4^2$

_____ [2]

Grade 4–6 **6**   A green light flashes every 12 seconds. A blue light flashes every 16 seconds.

After how many seconds will both lights flash together?

_____ [3]

Grade 4–6 **7**   A shop makes sandwiches on white, brown or granary bread. There are 6 different choices of fillings.

Show that there are 18 different sandwich types.

[1]

Grade 4–6 **8**   How many different ways can the letters A, B, C and D be arranged?

_____ [2]

Total Marks _____ / 15

Basic Algebra

 1 Expand and simplify $6(x + 4) - 5(3 - 2x)$

_____ [2]

 2 Factorise $8x^2 - 2x$

_____ [2]

 3 Expand and simplify $2x(3x + 1) + 5(3x - 2)$

_____ [2]

 4 Solve $2x + 4 = x + 1$

$x =$ _____ [2]

 5 One side of a square has length $(5x - 2)$ cm.

Work out the perimeter when $x = 3$

$(5x - 2)$ cm

_____ cm [3]

6 Solve $\frac{2x}{7} + 5 = 3$

$x =$ _____ [3]

7 Solve $\frac{4 + 3x}{2} = x - 5$

$x =$ _____ [3]

Total Marks _____ / 17

Factorisation and Formulae

Grade 4–6

1 Simplify $\dfrac{6(x-2)^2}{2(x-2)}$

_____ [2]

Grade 4–6

2 Expand and simplify $(4x+1)(2x-3)$

_____ [2]

Grade 4–6

3 Factorise $x^2 - x - 20$

_____ [2]

Grade 4–6

4 Factorise fully $4x^2 - 100$

_____ [2]

Grade 4–6

5 **a)** Make a the subject of the formula $v^2 = u^2 + 2as$

_____ [2]

Grade 4–6

b) Work out the value of a when $v = 8$, $u = 5$ and $s = 3$

$a = $ _____ [2]

Grade 4–6

6 Make A the subject of the formula $r = \sqrt{\dfrac{A}{\pi}}$

_____ [2]

Grade 7–9

7 Show that $\dfrac{12x - 60}{x(x^2 - 8x + 15)} \times \dfrac{x^2 - 3x}{3}$ simplifies to c where c is an integer.

[4]

Total Marks _____ / 18

Topic-Based Questions

7

Ratio and Proportion

Grade 4–6

1 A woman invests £1650 in shares and cash in the ratio 6 : 5.

How much does she invest in shares?

£ _____ [2]

Grade 4–6

2 A house is 14.4 metres high. A model is made using the scale 1 : 18.

Work out the height of the model. Give your answer in centimetres.

_____ cm [2]

Grade 4–6

3 There are 96 people on an aircraft. Half of them are men. 12 are children.

Work out the ratio in its simplest form   number of women : number of children

_____ [3]

Grade 4–6

4 A tap leaks 45 litres of water over 3 days. How much water will leak after 7 days?

State any assumptions you make.

_____ *l* [3]

Grade 7–9

5 Make a sketch on each graph to show the type of proportionality.

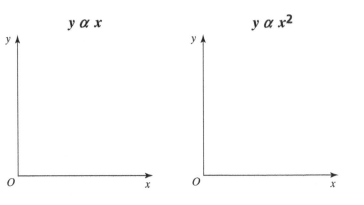

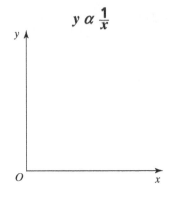

[3]

The graphs are labelled $y \alpha x$, $y \alpha x^2$, $y \alpha \frac{1}{x}$.

Total Marks _____ / 13

Variation and Compound Measures

 1 An aircraft flew 210 km in 1 hour 30 minutes. Work out the average speed.

_____ km/h [2]

 2 Matt drives at an average speed of 65 mph for 3 hours. How many miles does he drive?

_____ miles [2]

 3 The density of gold is 19.3 g/cm³. A ring has a mass of 4.5 grams and a volume of 0.3 cm³.

Is the ring made of pure gold? Use the formula $\text{Density} = \dfrac{\text{mass}}{\text{volume}}$

_____ [2]

 4 A bank has two compound accounts, Saver and Money Maker.

| **Saver**: Interest 1.5% per annum | **Money Maker**: Interest 2% per annum |

I invest £2000 in Saver and £1500 in Money Maker.

Work out which account will pay most interest after 2 years.

_____ [4]

 5 Here is a distance–time graph of a car. Work out the speed of the car in miles per hour.

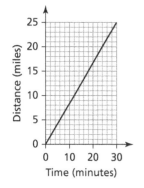

_____ mph [2]

Total Marks _____ / 12

Topic-Based Questions

9

Angles and Shapes 1 & 2

Grade 4–6   **1**   The diagram shows a kite. Work out the size of angle x.

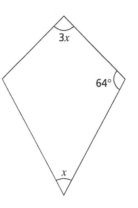

_____ °   **[3]**

Grade 4–6   **2**   The exterior angle of a regular polygon is 18°. Work out the number of sides of the polygon.

_____   **[2]**

Grade 4–6   **3**   The diagram shows an isosceles triangle touching two parallel lines.

Work out the size of angle x.

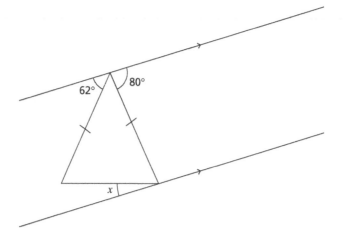

_____ °   **[4]**

Grade 4–6   **4**

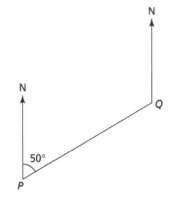

a)   Write down the bearing of Q from P.

_____ °   **[1]**

b)   Work out the bearing of P from Q.

_____ °   **[2]**

Total Marks _____ / 12

Fractions

Grade 4–6 **1** Find the number that is exactly halfway between $\frac{1}{3}$ and $\frac{5}{6}$

_____ [2]

Grade 4–6 **2** Work out $3\frac{1}{2} + 2\frac{2}{3}$

_____ [2]

Grade 4–6 **3** Work out $\frac{3}{4} \times \frac{2}{5}$. Give your answer as a fraction in its simplest form.

_____ [2]

Grade 4–6 **4** There are 8 boys in a classroom. This is $\frac{2}{5}$ of the class. How many students are in the class altogether?

_____ [3]

Grade 4–6 **5** Which is bigger, $\frac{5}{8}$ of 40 or $\frac{2}{3}$ of 36? You must show your working.

_____ [2]

Grade 4–6 **6** Write $0.\dot{7}$ as a fraction.

_____ [1]

Grade 4–6 **7** 180 people attend a party. $\frac{1}{3}$ are men, $\frac{1}{2}$ are women. What fraction are children?

_____ [2]

Grade 7–9 **8** Prove algebraically that $0.3\dot{1}\dot{5}$ can be written as $\frac{52}{165}$

_____ [3]

Total Marks _____ / 17

Percentages

Grade 4-6 **1** A coffee machine costs £800. VAT is added at 20%.

The machine is paid for in 12 equal payments. Work out the cost of each payment.

£ _____ [3]

Grade 4-6 **2** In a sale, the cost of a jacket is reduced from £150 to £105. Work out the percentage reduction.

_____ % [2]

Grade 4-6 **3** Jack wants to buy 30kg of pet food. He sees these two offers:

A	£23.50 for 15kg bag Offer 10% off

B	£8.50 for 5kg bag Buy 5 bags get one free

Which is the better offer for Jack?

_____ [4]

Grade 4-6 **4** In August, a clothes shop increases the price of shirts by 20%.

In September, the price is reduced by 5%.

Work out the overall percentage increase in price.

_____ % [3]

Grade 4-6 **5** A bed is in a sale with 20% off the normal price. The price is then further reduced by 5% of the sale price. The price is now £1235.

Work out the normal price.

£ _____ [3]

Total Marks _____ / 15

Probability 1 & 2

Grade 4–6 **1** A four-sided spinner is labelled A, B, C and D.

The table shows the probabilities of the spinner landing on A and on B.

The probability of landing on D is twice the probability of landing on C.

Letter	A	B	C	D
Probability	0.3	0.1		

a) Complete the table. [2]

b) The spinner is spun 50 times. Estimate the number of times it lands on A.

_____ [2]

Grade 4–6 **2** A bag contains three different shapes: circles, triangles and squares.

The probability of picking a circle at random is 0.1

There are twice as many triangles as squares.

a) Work out the probability of picking a triangle at random.

_____ [2]

b) There are 12 circles in the bag. Work out the number of squares in the bag.

_____ [2]

Grade 4–6 **3** There are 7 blue counters and 3 red counters in a bag. A counter is picked at random and replaced.

a) Write down the probability that the counter picked is blue.

_____ [1]

b) A second counter is picked. Work out the probability that both are red.

_____ [2]

4 ξ = {Integers from 11 to 20 inclusive}

A = {prime numbers}   B = {13, 14, 15, 16, 17}

a) Complete the Venn diagram.

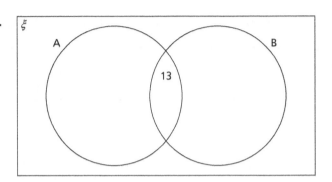

[3]

b) A number is chosen at random from ξ.

Work out the probability that the number is in set A $\cup$ B.

_____ [2]

5 A box contains 20 counters. 5 of the counters are red.

Two counters are taken out at random.

a) Complete the tree diagram.

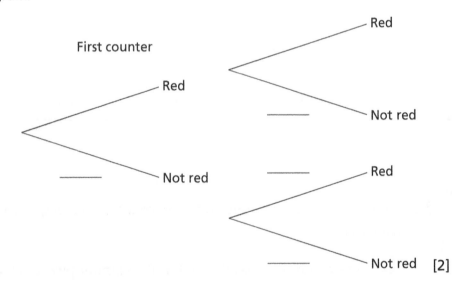

[2]

b) Work out the probability of picking two red counters.

_____ [2]

Total Marks _____ / 20

Number Patterns and Sequences & Terms and Rules

1 Work out the missing term of this geometric sequence.

3   6   24   48

_____ [1]

2 Here are the first four terms of an arithmetic sequence: 10   14   18   22

Write down an expression, in terms of n, for the nth term of the sequence.

_____ [2]

3 The nth term of a sequence is given by $an^2 + bn$, where a and b are integers.

The first term of the sequence is 1. The fourth term of the sequence is 28.

Work out the second and third terms of the sequence.

_____ [4]

4 The nth term of a linear sequence is $20 - 3n$

The nth term of a quadratic sequence is $n^2 + 5$

Show that there is only one number common to both sequences.

_____

_____ [3]

5 Here are the first five terms of a sequence: 0   3   10   21   36

Find an expression, in terms of n, for the nth term of this sequence.

_____ [3]

Total Marks _____ / 13

Transformations

1

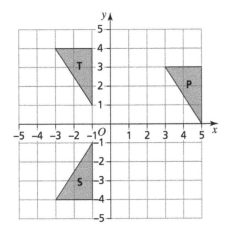

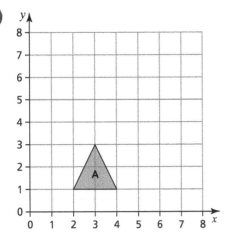

a) Rotate shape **T** 180° about the point (0, 1).

Label your shape **R**. [2]

b) Describe fully the single transformation that maps shape **T** onto shape **S**.

_____

_____ [2]

c) Shape **T** can be transformed to shape **P** using a translation $\begin{pmatrix} a \\ b \end{pmatrix}$.

Write down the values of a and b.

$a =$ _____

$b =$ _____ [2]

2

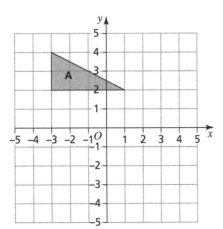

a) Reflect shape **A** in the line $y = 4$. Label it **B**. [2]

b) Enlarge shape **A** scale factor 2, centre of enlargement (0, 0). Label it **C**. [2]

3

Triangle **A** is reflected in the line $y = x$ to give triangle **B**.

Triangle **B** is reflected in the line $x = 1$ to give triangle **C**.

Describe the **single** transformation that will map triangle **C** to triangle **A**.

_____

_____ [3]

Total Marks _____ / 13

Constructions

Grade 4–6 **1** Use a ruler and compasses to construct the perpendicular bisector of the line *AB*.

[2]

Grade 4–6 **2** Use a ruler and compasses to construct the angle bisector of the angle C.

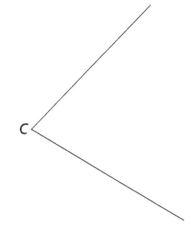

[2]

Grade 4–6 **3** Draw the locus of the points that are 2 cm from the line.

[2]

Total Marks .... / 6

Linear Graphs

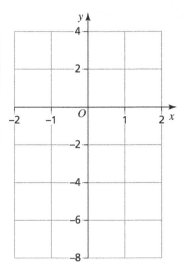

Grade 4–6

1 **a)** Complete the table of values for $y = 3x - 2$.   [2]

x	–2	–1	0	1	2
y					

b) On the grid, draw the graph of $y = 3x - 2$ for values of x from –2 to 2.   [2]

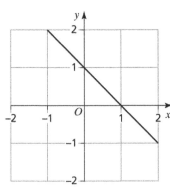

Grade 4–6

2 Here are six equations. $y = 1 - x$ $y = x - 1$ $y + 1 = 0$ $y = 1$ $x = 1$ $y = x + 1$

Match each graph to its equation.

A   **B**   **C**

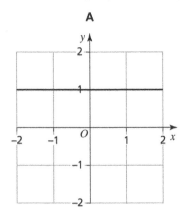 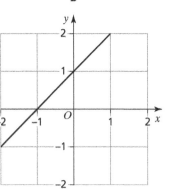 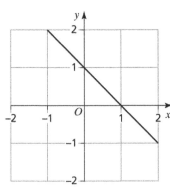

_____   _____   _____

[3]

Grade 4–6

3 A is the point (1, 2). B is the point (3, 8).

a) Work out the gradient of AB.

_____ [2]

b) Work out the equation of the line AB.

_____ [2]

c) C is the point (7, 20). Show that ABC is a straight line.

_____ [1]

Total Marks _____ / 12

Graphs of Quadratic Functions

1 a) Complete the table of values for $y = x^2 - x + 2$

x	–2	–1	0	1	2
y			2		4

[2]

b) On the grid, draw the graph of $y = x^2 - x + 2$ for values of x from –2 to 2.

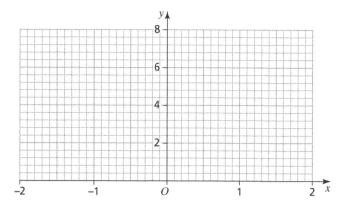

[2]

c) Use your graph to estimate solutions to $x^2 - x + 2 = 3$

$x =$ _____ , $x =$ _____   [2]

2 Here is the graph of $y = x^2 - 2x - 6$

a) Write down the coordinates of the turning point.

(_____ , _____)   [1]

b) Use the graph to find the roots of $x^2 - 2x - 6 = 0$

$x =$ _____ , $x =$ _____   [2]

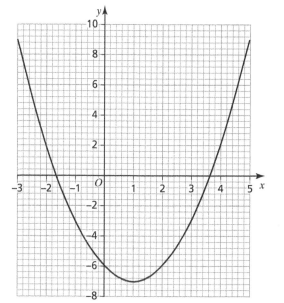

3 Shown below is a sketch of $y = f(x)$.

On the grid sketch the graph of $y = -f(x)$.   [1]

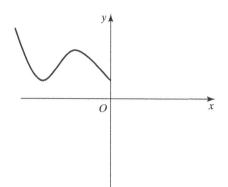

Total Marks _____ / 10

Topic-Based Questions   19

Powers, Roots and Indices

Grade 4–6 **1** Write down the value of $25^{\frac{1}{2}}$

_____ [1]

Grade 4–6 **2** Work out the value of $14^0 + 7^{-2}$

_____ [2]

Grade 4–6 **3** a) Simplify $(4ab^2)^3$

_____ [2]

b) Simplify $\dfrac{9x^6y^5}{3xy^3}$

_____ [2]

Grade 4–6 **4** Ali works out $d^4 \div d^4$. His answer is d.

Is he correct? Show working to support your answer.

_____ [1]

Grade 4–6 **5** Simplify fully $\dfrac{c^8 \times c^{-4}}{c^3}$

_____ [2]

Grade 4–6 **6** Solve $4^x = 64$

$x =$ _____ [1]

Grade 7–9 **7** Write down the value of $8^{-\frac{2}{3}}$

_____ [2]

Grade 7–9 **8** Rationalise the denominator of $\dfrac{10}{\sqrt{5}}$ Give your answer in its simplest form.

_____ [2]

Grade 7–9 **9** Write $\dfrac{\sqrt{3}}{\sqrt{3}-1}$ in the form $\dfrac{a+\sqrt{3}}{b}$ where a and b are integers.

_____ [2]

Total Marks _____ / 17

Area and Volume 1 & 2

Grade 4–6

1 The trapezium is the uniform cross-section of the prism of length 30 m.

Work out the volume of the prism.

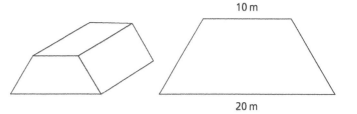

10 m

8 m

20 m

_____ m³   [3]

Grade 4–6

2 1 litre = 1000 cm³

This cylindrical tank is half filled with water.

How much water is in the tank? Give your answer in litres to 1 decimal place.

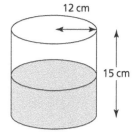

12 cm

15 cm

_____ *l*   [4]

Grade 7–9

3 The diagram shows two spheres. The ratio of the radius of the smaller sphere to the radius of the larger sphere is 2 : 3

Work out the surface area of the larger sphere.

4.8 cm

| Surface area of a sphere = $4\pi r^2$ |

_____ cm²   [3]

Grade 7–9

4 The diagram shows a cube and a square-based pyramid.

They both have the same volume. Work out the height of the pyramid.

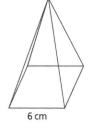

| Volume of a pyramid = $\frac{1}{3}$ × area of base × perpendicular height |

4.5 cm   6 cm

_____ cm   [3]

⟨ **Total Marks** _____ / 13 ⟩

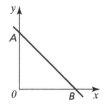
Uses of Graphs

1 Find the equation of the line that is parallel to the line $y = 5x + 1$ and passes through the point (0, –3).

_____ [2]

2 The graph shows a sketch of the line $3y + 4x = 12$. The line crosses the axes at A and B.

Work out the area of triangle OAB.

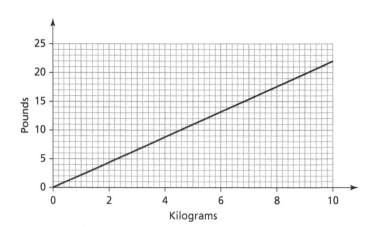

_____ units² [3]

3 You can use this graph to change between kilograms and pounds.

Change 60 pounds to kilograms.

_____ kg [2]

4 Line **L** has equation $2y = 3x - 4$

Line **M** is perpendicular to line **L** and passes through the point (0, 5).

Work out the equation of line **M** in the form $ax + by + c = 0$.

_____ [4]

Total Marks _____ / 11

Other Graphs 1 & 2

Grade 4–6

1 The graph shows part of a journey.

a) Calculate the distance travelled during this 20-second period.

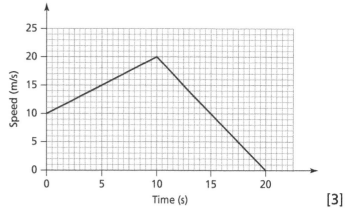

_____ m

[3]

b) Compare the acceleration and the deceleration in the two parts of the journey.

_____

[1]

Grade 7–9

2 The graph shows part of a journey.

Work out an estimate of the speed at time 3 seconds.

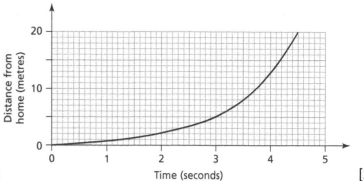

_____ m/s

[3]

Grade 7–9

3 On the grid sketch the graph of $x^2 + y^2 = 4$

Label the points of intersection with the axes.

[2]

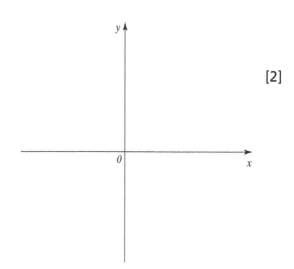

4 The graph shows the speed of a bus during the first 20 seconds after leaving a bus stop.

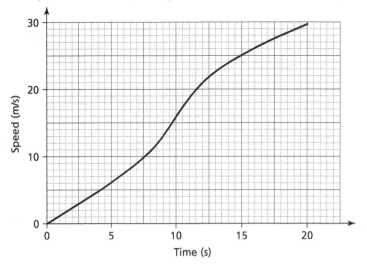

a) Calculate an estimate of the gradient of the graph at 15 seconds.

_____ [3]

b) Describe what this answer represents.

_____ [1]

c) Work out an estimate for the distance travelled in the 20 seconds shown. Use four strips of equal width.

_____ m [3]

5 Here are four equations. $y = x^3$ $y = 3^x$ $y = \dfrac{3}{x}$ $y = 3x$

Three of the graphs are sketched. Match each sketch graph to its equation.

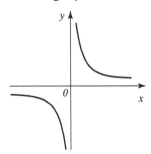 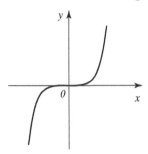 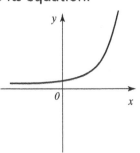

_____   _____   _____ [3]

Inequalities

Grade 4–6 **1** On the number line, show the set of values for x for which $-2 < x - 1 \leqslant 2$

[2]

Grade 4–6 **2** Solve $4x - 7 \geqslant x + 8$

_____ [2]

Grade 4–6 **3** $-13 \leqslant 5x + 2 \leqslant 12$. Work out the greatest possible value of x^2.

_____ [3]

Grade 7–9 **4** Solve $(x + 1)^2 > 9$

_____ [3]

Grade 7–9 **5** n is an integer such that $\dfrac{n}{n^2 + 12} > \dfrac{1}{8}$

Find all possible values of n.

_____ [4]

Grade 7–9 **6** Solve $2 < \dfrac{x^2 - 1}{12} < 4$

_____ [4]

Total Marks _____ / 18

Congruence and Geometrical Problems

Grade 4–6

1 Triangle *ABC* and triangle *XYZ* are similar.

a) Work out the length of *BC*.

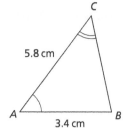

_____ cm [2]

b) Work out the perimeter of triangle *XYZ*.

_____ cm [3]

Grade 7–9

2 Here are two similar triangles. Work out the value of x.

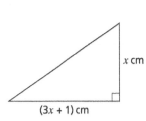
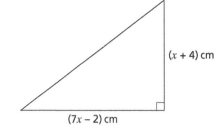

$x =$ _____ [5]

Grade 7–9

3 In the diagram *AB* is parallel to *DC* and *AD* is parallel to *BC*.

Prove that triangle *ABD* is congruent to triangle *CDB*.

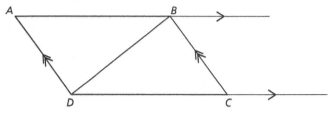

_____

_____

_____ [4]

Total Marks _____ / 14

Right-Angled Triangles

Grade 4–6

1 *ABC* is a right-angled triangle.

Work out the length of *BC*.

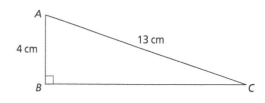

_____ cm   [3]

2 The diagram shows two congruent triangles and a square.

Work out the area of the square.

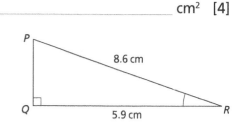

_____ cm²   [4]

3 *PQR* is a right-angled triangle.

Work out the size of angle *PRQ*.

Give your answer to 1 decimal place.

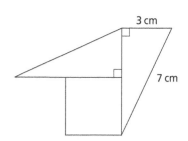

_____ °   [2]

4 Here are two right-angled triangles:

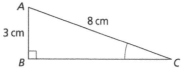

Which angle is bigger, *ACB* or *XZY*? You must give a reason for your answer.

_____

_____   [1]

5 The diagram shows a triangular prism.

Work out its volume.
Give your answer correct to 3 significant figures.

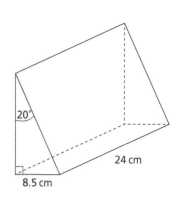

_____ cm³   [4]

Total Marks _____ / 14

Sine and Cosine Rules

Grade 7–9 **1** Work out the size of angle *BAC*.

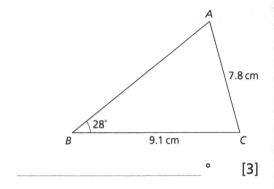

_____ ° **[3]**

Grade 7–9 **2** In the diagram *BCD* is a straight line.

a) Work out the length of *BD*.

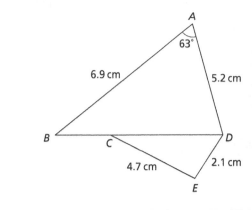

_____ cm **[3]**

b) *BC* : *CD* = 2 : 3

Work out the size of angle *CED*.

_____ ° **[4]**

Grade 7–9 **3** The diagram shows a cuboid *ABCDEFGH*.

Work out the size of angle *BEC*.

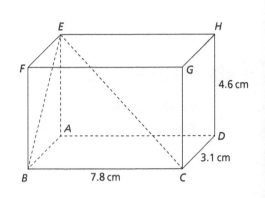

_____ ° **[5]**

Total Marks _____ / 15

Statistics 1

Grade 4–6 **1** The scatter graph shows information about the marks of some students in two tests.

a) Write down the type of correlation.

_____ [1]

b) One student had a mark that was an outlier.

Circle the cross for that student on the graph. [1]

c) Leo got a mark of 21 on Test A.

Use the scatter graph to estimate his mark on Test B.

_____ [2]

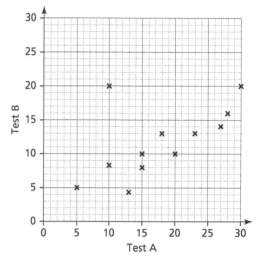

Grade 4–6 **2** The table shows information about the times taken, in minutes, by 20 students to complete a task.

a) Find the class interval that contains the median.

_____ [2]

Time (t minutes)	Frequency
$0 < t \leq 2$	4
$2 < t \leq 4$	5
$4 < t \leq 6$	8
$6 < t \leq 8$	3

[2]

b) Work out an estimate of the mean time.

_____ [3]

Grade 4–6 **3** The stem and leaf diagram shows the ages, in years, of the 19 workers in a factory.

```
1 | 7  9  9
2 | 1  4  4  6  6  6  7  9
3 | 2  2  3  7  7  8
4 | 3  5
```

Key: 1 | 7 represents age 17

Complete the table.

Least age	Lower quartile	Median	Upper quartile	Greatest age
17				45

[2]

Total Marks _____ / 11

Grade 4–6

1 Here is some information about the heights of 60 plants.

Least height	Lower quartile	Median	Upper quartile	Greatest height
8 cm	15 cm	20 cm	29 cm	42 cm

a) Draw a box plot to represent this information.

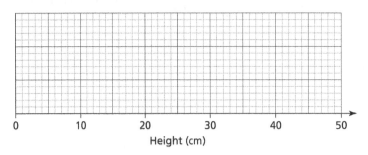

Height (cm)

[3]

b) Work out an estimate for the number of plants with a height greater than 15 cm.

_____ [2]

Grade 7–9

2 The histogram shows some information about the heights of buildings on a street.

The two bars shown represent one-third of the buildings.

a) Complete the histogram. [4]

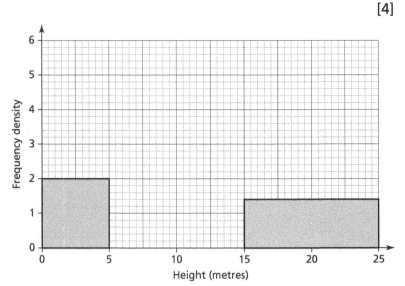

Height (metres)

b) Estimate the median height.

_____ m   [2]

Total Marks _____ / 11

Grade 4–6

1 The length of a pencil is 17.6 cm to 1 decimal place.

Complete the error interval for the length of the pencil.

_____ cm ⩽ length < _____ cm   [2]

Grade 4–6

2 Naby changes £150 into US dollars. The exchange rate is £1 = $1.38

a) Does he have enough dollars to buy jeans costing $70 and a jacket costing $125?

You must show your working.

_____   [3]

b) The same jeans cost £59 in England.

Are the jeans cheaper in England or the USA? You must show your working.

_____   [2]

Grade 4–6

3 a) Use approximations to estimate the value of $\dfrac{199.4 \times 9.8}{50.2}$

_____   [3]

b) Is your answer to part **a)** an underestimate or an overestimate?

Give a reason for your answer.

_____

_____   [1]

Grade 7–9

4 The radius of a circle is $r = 5.36$ cm correct to 3 significant figures.

By considering bounds, work out the area of the circle to a suitable degree of accuracy.

Give a reason for your answer.

_____

_____

_____   [4]

Total Marks _____ / 15

Solving Non-Linear Equations

Grade 4–6 **1** Solve $x^2 - 2x - 15 = 0$

$x =$ _____ or $x =$ _____ [2]

Grade 4–6 **2** The area of this rectangle is $84\,cm^2$.

a) Form an equation to show this information.

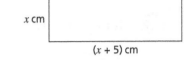

x cm

$(x + 5)$ cm

_____ [1]

b) Solve the equation to work out the value of x.

$x =$ _____ [3]

Grade 7-9 **3** Given that $x^2 - 8x + 2 = (x - a)^2 - b$ for all values of x.

a) Find the value of a and the value of b.

$a =$ _____ , $b =$ _____ [2]

b) Hence write down the coordinates of the turning point on the graph of $y = x^2 - 8x + 2$

(_____ , _____) [1]

Grade 7-9 **4** a) Show that the equation $x^3 + 2x = 20$ has a solution between 2 and 3.

_____

_____ [2]

b) Show that the equation $x^3 + 2x = 20$ can be rearranged to give $x = \sqrt[3]{20 - 2x}$

_____

_____ [1]

c) Starting with $x_0 = 2$, use the iteration formula $x_{n+1} = \sqrt[3]{20 - 2x_n}$ to estimate a solution to $x^3 + 2x = 20$

_____ [3]

Total Marks _____ / 15

Simultaneous Equations and Functions

Grade 4–6

1 Solve the simultaneous equations $2x + y = 1$

$$x - 2y = 8$$

$x =$ _____ , $y =$ _____ [4]

Grade 7–9

2 Solve algebraically the simultaneous equations $5x^2 - y^2 = 4$

$$3x + 2y = 2$$

$x =$ _____ , $y =$ _____ or $x =$ _____ , $y =$ _____ [5]

Grade 7–9

3 f and g are functions such that $f(x) = x^2 + 3$ and $g(x) = 4x - 1$

a) Find f(–4)

_____ [1]

b) Find fg(5)

_____ [2]

Grade 7–9

4 The function f is given by $f(x) = 5x^2 + 1$

a) Show that if $x > 0$, $f^{-1}(6) = 1$

_____

_____

_____ [2]

b) $g(x) = 2x$

Solve the equation $gf(x) = 3(x + 1)$

$x =$ _____ or $x =$ _____ [4]

Total Marks _____ / 18

Algebraic Proof

1 Prove that the sum of the squares of any two consecutive odd numbers is a multiple of 2.

[3]

2 Prove that the difference of the squares of any two consecutive even numbers is a multiple of 4.

[3]

3 Prove that the sum of any five consecutive numbers is a multiple of 5.

[2]

4 Show that $\dfrac{1}{x+1} - \dfrac{4}{x^2+6x+5}$ simplifies to $\dfrac{1}{x+5}$

[4]

5 a) Show that $\dfrac{5}{x-1} + \dfrac{2}{x+1} = 1$ simplifies to $x^2 - 7x - 4 = 0$

[4]

b) Hence solve the equation $\quad \dfrac{5}{x-1} + \dfrac{2}{x+1} = 1$

Give your answers as surds.

[2]

Total Marks .... / 18

Circles

1 ABCD is a trapezium with AD parallel to BC.

O is the centre of the circle.

Work out the size of angle ADB.

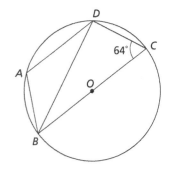

_____ ° [3]

2 O is the centre of the circle. RTS is a tangent to the circle at T.

ROP is a straight line with P on the circle.

Work out the size of angle PTS.

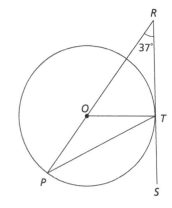

_____ ° [4]

3 Triangle BCD is isosceles. BE is a tangent to the circle.

Prove that angle BAC = 2x

_____

_____

_____

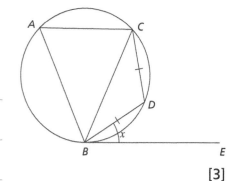

[3]

4 In the diagram O is the centre of the circle.

Work out the size of angle CAB.

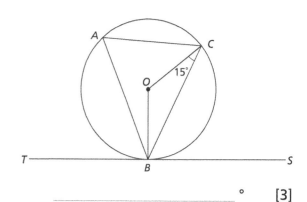

_____ ° [3]

Total Marks _____ / 13

Vectors

1 $a = \begin{pmatrix} 2 \\ -7 \end{pmatrix}$ $b = \begin{pmatrix} 3 \\ 4 \end{pmatrix}$

Work out **3a + 4b** as a column vector.

$\begin{pmatrix} \\ \end{pmatrix}$ [2]

2 *OACB* is a parallelogram. $\overrightarrow{OA} = $ **a**, $\overrightarrow{OB} = $ **b**

P is a point on OC such that OP : PC = 1 : 2

M is the midpoint of OB.

a) Work out $\overrightarrow{OP}$ in terms of **a** and **b**.

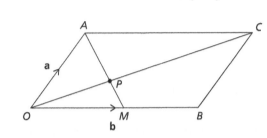

_____ [2]

b) Work out $\overrightarrow{AP}$ in terms of **a** and **b**.

_____ [2]

c) Work out the ratio AP : PM

_____ [3]

3 *OAB* is a triangle. $\overrightarrow{OA} = $ **a**, $\overrightarrow{OB} = $ **b**

OT : TA = 1 : 4   AS : SB = 5 : 2

a) Write down $\overrightarrow{TA}$ in terms of **a**.

_____ [1]

b) Work out $\overrightarrow{TS}$ in terms of **a** and **b**.

Simplify your answer.

_____ [4]

Total Marks _____ / 14

Collins

GCSE
Mathematics
Paper 1 Higher Tier (Non-Calculator)

H

Time: 1 hour 30 minutes

You must have:

- Ruler graduated in centimetres and millimetres, protractor, pair of compasses, pen, HB pencil, eraser.

You may not use a calculator

Instructions

- Use **black** ink or black ball-point pen.
- Answer **all** questions.
- Answer the questions in the spaces provided – *there may be more space than you need.*
- **Calculators may not be used.**
- Diagrams are NOT accurately drawn, unless otherwise indicated.
- You must **show all your working out**.

Information

- The total mark for this paper is 80.
- The marks for **each** question are shown in brackets
  - *use this as a guide as to how much time to spend on each question.*
- Read each question carefully before you start to answer it.
- Keep an eye on the time.
- Try to answer every question.
- Check your answers if you have time at the end.

Name: ....

Answer ALL questions.

Write your answers in the spaces provided.

You must write down all stages in your working.

1   Sam drives 162 miles from Sheffield to London on the motorway.
    50 of the miles have roadworks.

   (a) Work out the journey time.
       Assume an average speed of: 50 mph through the roadworks

       70 mph for the rest of the journey.

                                                                        _____ (4)

   (b) Sam arrives later than expected.

       What does this tell you about the average speed?

       _____ (1)

                                                                    (Total for Question 1 is 5 marks)

2   Here is a formula: $V = 3r^2h$

   Work out the value of V when $r = -4$ and $h = 2$

                                                   $V =$ _____

                                                                    (Total for Question 2 is 2 marks)

3   Work out an estimate for the value of $\dfrac{139.8 \times 1.9}{70.7}$

_____

(Total for Question 3 is 3 marks)

4   The diagram shows a regular pentagon made from five identical triangles.

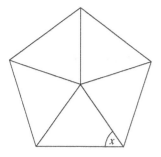

Show that $x = 54°$

(Total for Question 4 is 3 marks)

Turn over

5 Box A contains 4 blue pencils and 6 green pencils.

Box B contains 6 blue pencils and 10 green pencils.

A pencil is taken at random from each box.

Ali says, "I am more likely to choose a blue pencil from Box B because there are more in that box."

Is he correct?

You must show your working.

_____

(Total for Question 5 is 3 marks)

6 **(a)** Solve $4(x - 5) = 24$

$x = $ _____ **(3)**

(b) Factorise fully $8x^2 - 12x$

_____ **(2)**

(Total for Question 6 is 5 marks)

7   (a) Write 237 000 in standard form.

_____   (1)

(b) Write 4.5×10^{-4} as an ordinary number.

_____   (1)

(c) Which of the following two numbers is greater?

$$8.91 \times 10^3 \qquad \text{or} \qquad 5.62 \times 10^4$$

Give a reason for your answer.

_____   (1)

(Total for Question 7 is 3 marks)

8   In a school, the ratio of boys to girls is 3 : 2
25% of the boys study French.
50% of the girls study French.

What percentage of the students study French?

_____ %

(Total for Question 8 is 3 marks)

Turn over

9 Solve $x^2 - 11x + 28 = 0$

$x =$ _____ or $x =$ _____

(Total for Question 9 is 3 marks)

10 $\mathbf{a} = \begin{pmatrix} 4 \\ 9 \end{pmatrix}$ $\mathbf{b} = \begin{pmatrix} 3 \\ -2 \end{pmatrix}$

Work out $\mathbf{a} + 2\mathbf{b}$ as a column vector.

$\begin{pmatrix} \underline{} \\ \underline{} \end{pmatrix}$

(Total for Question 10 is 2 marks)

11 *A, B, C* and *D* are points on the circumference of a circle.

TBS is a tangent to the circle.

Angle *ADB* = 32°

Angle *BCD* = 66°

Triangle *BCD* is isosceles.

Work out the size of angle *ABC*.

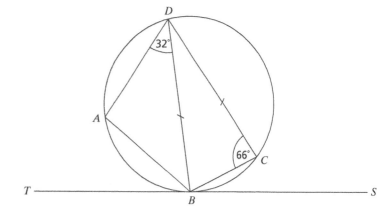

.... °

(Total for Question 11 is 4 marks)

Turn over

12   The table shows the ages of 100 people in a small village.

Age (A years)	Frequency
$0 < A \leqslant 20$	15
$20 < A \leqslant 40$	36
$40 < A \leqslant 60$	32
$60 < A \leqslant 80$	17

(a) On the grid below, draw a cumulative frequency graph for this information.

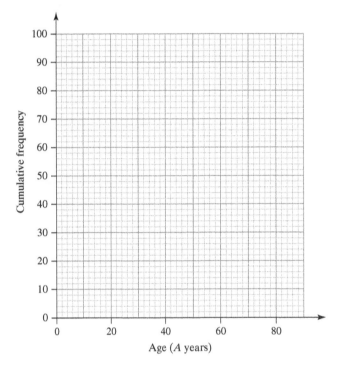

(2)

(b) Find an estimate of the upper quartile of the ages of the 100 people.

_____ years   (2)

(Total for Question 12 is 4 marks)

13 **(a)** $(\sqrt{3} + 1)(\sqrt{6} - \sqrt{2})$ can be written in the form $a\sqrt{2}$ where a is an integer.

Find the value of a.

$a =$ _____ (3)

(b) Rationalise the denominator of $\dfrac{10}{(3 + \sqrt{5})}$

Give your answer in its simplest form.

_____ (3)

(Total for Question 13 is 6 marks)

14 Prove algebraically that $0.1\dot{7}$ can be written as $\dfrac{8}{45}$

(Total for Question 14 is 3 marks)

Turn over

15 **(a)** Work out the value of $\left(\dfrac{8}{125}\right)^{\frac{2}{3}}$

_____ (2)

(b) Solve $3^x = \dfrac{1}{81}$

$x =$ _____ (1)

(Total for Question 15 is 3 marks)

16 y is directly proportional to x^2
When $x = 4$, $y = 48$

(a) Find a formula for y in terms of x.

_____ (3)

(b) Work out the positive value of x when $y = 27$

$x =$ _____ (2)

(Total for Question 16 is 5 marks)

17 The diagrams show a hemisphere of radius 3 cm and a cone with base radius 6 cm.

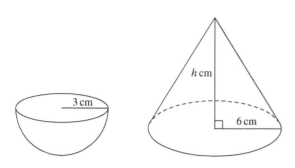

Volume of a cone = $\frac{1}{3}\pi r^2 h$

Volume of a sphere = $\frac{4}{3}\pi r^3$

The ratio of the volume of the hemisphere to the volume of the cone is 3 : 8

Work out the value of h.

$h =$ _____

(Total for Question 17 is 4 marks)

18 Show that the value of $\dfrac{1}{(\tan 30°)^2} \times \dfrac{1}{\cos 60°}$ is an integer.

_____

(Total for Question 18 is 2 marks)

Turn over

19 There are only yellow counters (y) and red counters (r) in a bag.

The ratio of yellow counters to red counters is $3 : 4$

A counter is taken out and then replaced.

6 yellow counters are then added to the bag.

The probability of taking out at random two yellow counters is $\dfrac{3}{14}$

Work out the number of red counters in the bag originally.

_____

(Total for Question 19 is 5 marks)

20 Solve the equation $\dfrac{3x^2}{4 - x} = 2$

$x =$ _____

(Total for Question 20 is 4 marks)

21   The graph of $y = f(x)$ is shown on the grid.

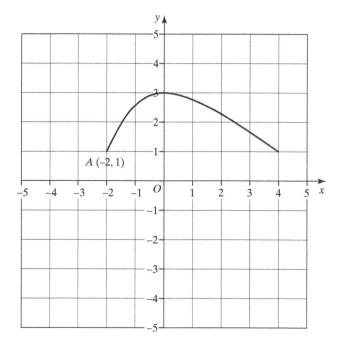

$A(-2, 1)$

(a) On the grid above, draw the graph of $y = -f(x)$

(1)

(b) When the graph of $y = f(x)$ is transformed to the graph with equation $f(x + 1) - 2$, point A is mapped to point B.

Write down the coordinates of point B.

(————— , —————)   (1)

(c) When the graph of $y = f(x)$ is reflected in the y-axis, one of the points is invariant.

Write down the coordinates of this point.

(————— , —————)   (1)

(Total for Question 21 is 3 marks)

Turn over

22 Sketch the graph of $y = 2x^2 - 8x - 1$

Show the coordinates of the turning point and the exact coordinates of any intercepts with the coordinate axes.

(Total for Question 22 is 5 marks)

TOTAL FOR PAPER IS 80 MARKS

Collins

GCSE
Mathematics
Paper 2 Higher Tier (Calculator)

H

Time: 1 hour 30 minutes

You must have:

- Ruler graduated in centimetres and millimetres, protractor, pair of compasses, pen, HB pencil, eraser, calculator.

Instructions

- Use **black** ink or black ball-point pen.
- Answer **all** questions.
- Answer the questions in the spaces provided – *there may be more space than you need.*
- **Calculators may be used.**
- If your calculator does not have a π button, take the value of π to be 3.142 unless the question instructs otherwise.
- Diagrams are NOT accurately drawn, unless otherwise indicated.
- You must **show all your working out**.

Information

- The total mark for this paper is 80.
- The marks for **each** question are shown in brackets
  - *use this as a guide as to how much time to spend on each question.*
- Read each question carefully before you start to answer it.
- Keep an eye on the time.
- Try to answer every question.
- Check your answers if you have time at the end.

Name: ....

Answer ALL questions.

Write your answers in the spaces provided.

You must write down all stages in your working.

1   Here is a list of numbers: 7 x $3x$ 6   9
     The mean is 40.

     Work out the value of x.

                                                            $x =$ _____

                                                   (Total for Question 1 is 3 marks)

2   **(a)** The density of copper is 9 grams/cm^3
         The volume of copper used to make a bracelet is 2.5 cm^3.

         Work out the mass of the bracelet.

                                            _____ g   (2)

     (b) A force of 50 Newtons acts on an area of 16 cm^2.

         $$\text{pressure} = \frac{\text{force}}{\text{area}}$$

         The force stays the same but the area increases.

         What happens to the pressure?

     _____   (1)

                                                   (Total for Question 2 is 3 marks)

3 **(a)** Find the highest common factor of 36 and 54.

.... (2)

(b) Find the lowest common multiple of 9, 12 and 18.

.... (2)

(Total for Question 3 is 4 marks)

4 Describe fully the single transformation that maps shape **A** onto shape **B**.

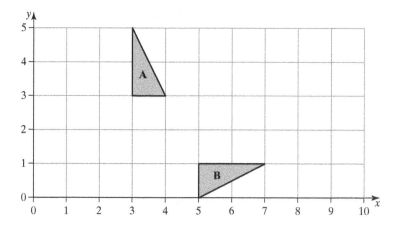

....

....

....

(Total for Question 4 is 3 marks)

Turn over

5   **(a)** Work out $\dfrac{(839 \times 74)}{\sqrt{97}}$

     Write down all the figures on your calculator display.

....   (2)

   (b) Write your answer to part (a) correct to 3 significant figures.

....   (1)

(Total for Question 5 is 3 marks)

6   These two triangles are similar:

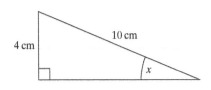 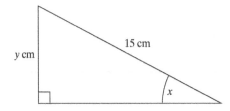

   Work out the value of y.

$y =$ ....

(Total for Question 6 is 2 marks)

7   The results when an ordinary, six-sided dice is rolled 60 times are shown.

Number shown	1	2	3	4	5	6
Frequency	5	8	12	15	9	11

(a) Do you think the dice is fair?
Give a reason for your answer.

_____

_____   (1)

(b) Work out the relative frequency of rolling a 5 or a 6.

_____   (2)

(Total for Question 7 is 3 marks)

8   A supermarket carries out an online survey of 50 customers.
15 customers say that they usually visit the supermarket on Fridays.

(a) On one Friday, approximately 4500 customers visit the supermarket.

Use this information to estimate how many customers the supermarket has altogether.

_____   (2)

(b) State any assumptions you made.

_____

_____   (1)

(Total for Question 8 is 3 marks)

Turn over

9   Here are two piles of exercise books.
     Each book is 0.75 cm thick.
     The smaller pile is 11.25 cm high.

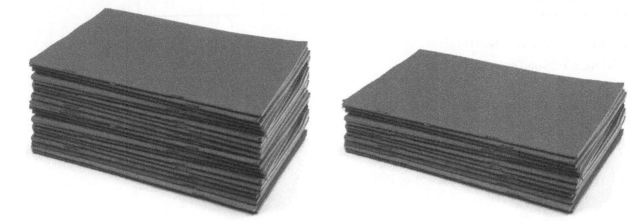

The height of the smaller pile is $\frac{3}{5}$ of the height of the taller pile.

Work out the total number of books in the two piles.

_____

(Total for Question 9 is 4 marks)

10   The length, L, of a runway is 1509 metres to the nearest metre.

     Write down the error interval for L.

_____ m $\leqslant L <$ _____ m

(Total for Question 10 is 2 marks)

11 f and g are functions such that $f(x) = 3x^2$ and $g(x) = \dfrac{5}{x}$

(a) Find f(–5)

_____ (1)

(b) Find fg(10)

_____ (2)

(c) $h(x) = 2 \sin x°$

Work out the value of $h^{-1}(0.4)$
Give your answer correct to 1 decimal place.

_____ (3)

(Total for Question 11 is 6 marks)

12 Expand and simplify $(4x + 3)(2x + 1)(x - 1)$

_____

(Total for Question 12 is 3 marks)

Turn over

13 Theo is making a toy out of cardboard as shown.

The outer shape is a sector of a circle of radius 4.5 cm.

The inner shape is a circle of radius 1 cm.

The angle between the radii is 25°

The circle is cut out.

Work out the area of toy remaining.

Give your answer correct to 3 significant figures.

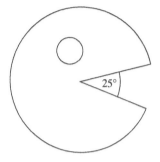

_____ cm²

(Total for Question 13 is 5 marks)

14 The graph shows the speed of a car along a test track.

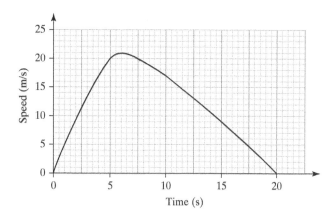

(a) Calculate an estimate of the gradient after 10 seconds.

_____ (3)

(b) Describe what the answer to part (a) represents.

_____ (1)

(c) Work out an estimate for the distance travelled in the first 10 seconds.
Use two strips of equal width.

_____ m (3)

(Total for Question 14 is 7 marks)

15 Prove algebraically that the sum of the squares of any two different odd numbers is always even.

(Total for Question 15 is 3 marks)

Turn over

16 The diagram shows two triangles ABD and ADC.

$BD : AC = 3 : 2$

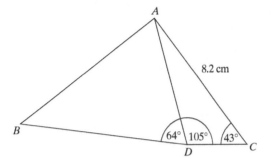

Work out the length of AB.

Give your answer to 3 significant figures.

_____ cm

(Total for Question 16 is 6 marks)

17 There are 24 teams in a competition.

Each team plays one match against each other.

Work out the number of matches played in the competition.

_____

(Total for Question 17 is 2 marks)

18 x and y are two numbers such that

$$x - 4 : y - 4 = 1 : 4$$

and $x + 6 : y + 6 = 1 : 2$

Work out the ratio $x : y$

_____ : _____

(Total for Question 18 is 5 marks)

Turn over

19 The diagram shows the types of shapes in a bag.
The shapes are circles, triangles and squares.
Each shape is lettered A or B.
The ratio of circles to triangles to squares is $2 : 3 : 4$
For each shape the ratio labelled A to B is $1 : 6$

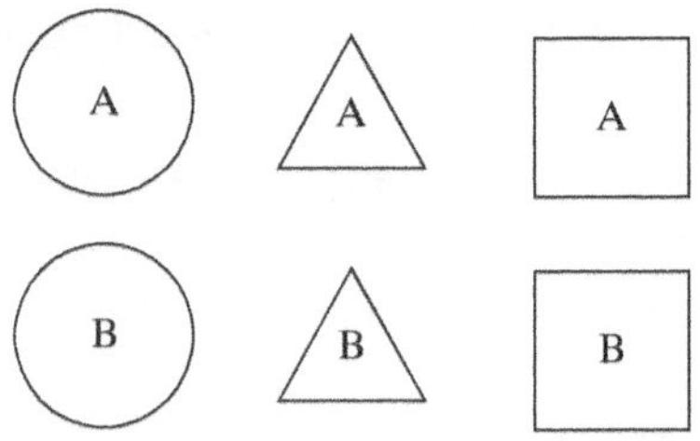

(a) Show that there must be more than 60 shapes in the bag.

(2)

(b) A shape is chosen at random.

Work out the probability it is a circle labelled A.

_____ (4)

(Total for Question 19 is 6 marks)

20   The diagram shows a triangle.

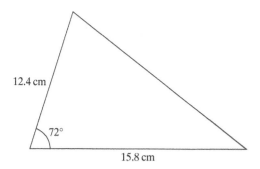

12.4 cm

72°

15.8 cm

Work out the area.

_____ cm²

(Total for Question 20 is 2 marks)

Turn over

21 The length of times that 60 students take to complete a task are shown in the table.

Time (t minutes)	Frequency
$0 < t \leqslant 1$	8
$1 < t \leqslant 4$	9
$4 < t \leqslant 6$	18
$6 < t \leqslant 10$	25

(a) Draw a histogram for the information in the table.

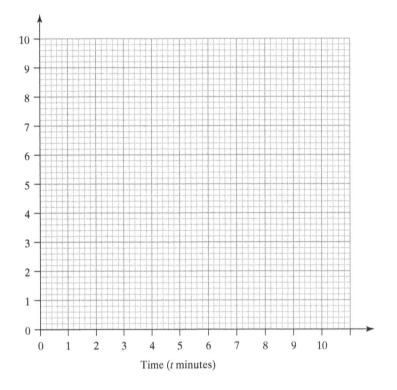

Time (t minutes)

(3)

(b) Estimate the median time.

_____ minutes   (2)

(Total for Question 21 is 5 marks)

TOTAL FOR PAPER IS 80 MARKS

Collins

GCSE
Mathematics

Paper 3 Higher Tier (Calculator)

H

Time: 1 hour 30 minutes

You must have:

- Ruler graduated in centimetres and millimetres, protractor, pair of compasses, pen, HB pencil, eraser, calculator.

Instructions

- Use **black** ink or black ball-point pen.
- Answer **all** questions.
- Answer the questions in the spaces provided – *there may be more space than you need*.
- **Calculators may be used.**
- If your calculator does not have a π button, take the value of π to be 3.142 unless the question instructs otherwise.
- Diagrams are NOT accurately drawn, unless otherwise indicated.
- You must **show all your working out**.

Information

- The total mark for this paper is 80.
- The marks for **each** question are shown in brackets
  - *use this as a guide as to how much time to spend on each question.*
- Read each question carefully before you start to answer it.
- Keep an eye on the time.
- Try to answer every question.
- Check your answers if you have time at the end.

Name: ....

Practice Exam Paper 3

Answer ALL questions.

Write your answers in the spaces provided.

You must write down all stages in your working.

1 *ABC* is a right-angled triangle.

Work out the length *BC*.
Give your answer correct to 1 decimal place.

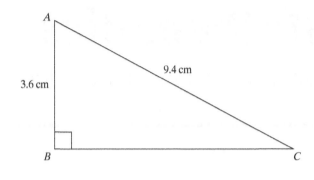

_____ cm

(Total for Question 1 is 3 marks)

2 Work out the value of $\dfrac{5^4 \times 5^6}{5^8 \times 5^{-1}}$

_____

(Total for Question 2 is 3 marks)

3   Three teams A, B and C play a game.

The probability that B wins to the probability that C wins is in the ratio 2 : 3

The probability that C wins = 0.54

Work out the probability that A wins.

_____

(Total for Question 3 is 4 marks)

4   The times taken by 60 people to complete a puzzle are shown.

Time (t minutes)	$0 < t \leqslant 5$	$5 < t \leqslant 10$	$10 < t \leqslant 15$	$15 < t \leqslant 20$
Frequency	22	14	13	11

(a) Find the interval that contains the median.

_____ (1)

(b) On the grid, draw a frequency polygon for the information in the table.

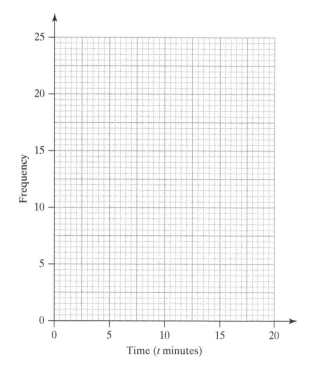

Time (t minutes)

(2)

(Total for Question 4 is 3 marks)

Turn over

5   1 gallon = 4.5 litres

The diagram shows a large empty water tank.

It holds 250 gallons when full.

Water is pumped into the tank at the rate of 12 litres per minute.

How long will it take to fill the tank?

Give your answer in hours and minutes to the nearest minute.

_____

(Total for Question 5 is 4 marks)

6   A fair spinner numbered from 1 to 5 is spun twice.

(a) Complete the probability tree diagram to show the probabilities of landing on even or odd numbers.

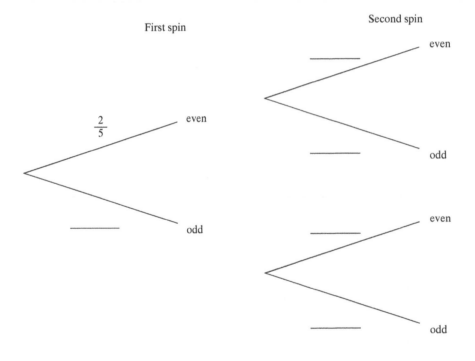

(2)

(b) Work out the probability of both spins landing on an even number.

(2)

_____

(Total for Question 6 is 4 marks)

7 Solve the simultaneous equations

$$3x + 2y = 11$$

$$x - 4y = 13$$

$x =$ _____

$y =$ _____

(Total for Question 7 is 3 marks)

Turn over

8   **(a)** Complete the table of values for $y = x^2 - 5x + 2$

x	–1	0	1	2	3	4	5	6
y		2			–4		2	

(2)

(b) On the grid, draw the graph of $y = x^2 - 5x + 2$ for values of x from –1 to 6.

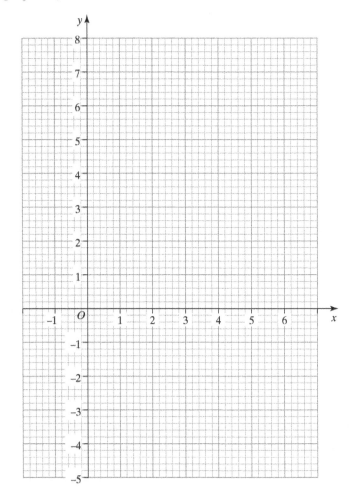

(2)

(c) Use your graph to estimate the coordinates of the turning point.

(———— , ————)   (2)

(Total for Question 8 is 6 marks)

9 A field has two parts, a sector of a circle and a parallelogram as shown.

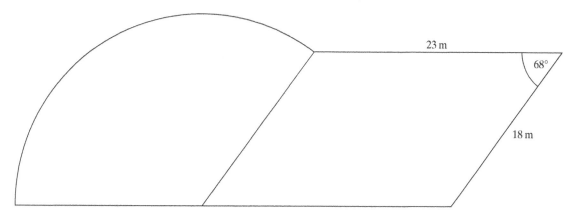

23 m

68°

18 m

Work out the perimeter of the field.
Give your answer to 3 significant figures.

_____ m

(Total for Question 9 is 5 marks)

10 The nth term of a sequence is given by the formula $2n^2 + c$ where c is an integer.
The 2nd term of the sequence is 9.

Work out the 7th term of the sequence.

_____

(Total for Question 10 is 3 marks)

Turn over

11 Anna invested £2780 in a savings account for five years.

She was paid 1.5% per annum compound interest for each of the first two years.

She was then paid 0.75% per annum compound interest for the last three years.

Anna said, "I will have more than £3000 in my account at the end of five years."

Is she correct?

You must show your working.

(Total for Question 11 is 3 marks)

12 A biased spinner has odd and even numbers on it.

When it is spun three times the probability of getting three even numbers is $\frac{27}{125}$

Work out the probability of getting two odd numbers if it is spun twice.

(Total for Question 12 is 2 marks)

13 The speed of sound (Mach 1) is 1.236×10^3 kilometres per hour.

A world record flight speed was set by a rocket which reached Mach 9.6 or $9.6 \times$ Mach 1.

Work out the distance that a rocket travelling at Mach 9.6 covers in 1 minute.

_____ km

(Total for Question 13 is 3 marks)

14 Make d the subject of the formula $\quad c = \dfrac{5d - 2}{d + 1}$

_____

(Total for Question 14 is 4 marks)

15 The diagram shows a tetrahedron of volume V.

> Volume of a tetrahedron $= \dfrac{1}{3} \times$ base area $\times$ height

If the area of the base, A, is increased by 25%
work out the percentage decrease in the height,
h, to maintain the same volume, V.

You must show your working.

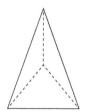

_____ %

(Total for Question 15 is 3 marks)

Turn over

16 The straight line L has the equation $2y + 5x - 10 = 0$
The line intercepts the x-axis at A and the y-axis at B.

Work out the equation of the perpendicular bisector of AB.
Give your answer in the form $ay + bx + c = 0$, where a, b and c are integers.

(Total for Question 16 is 5 marks)

17 The Venn diagram shows information about a jewellery collection.

$\xi = 80$ pieces of jewellery in the collection
G = jewellery containing some gold
S = jewellery containing some silver

$x : y = 6 : 1$ $\qquad$ $y : z = 3 : 10$

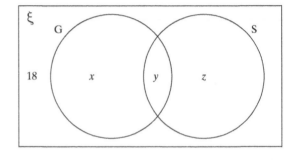

A piece of jewellery is chosen at random.
It is known to contain gold or silver or both gold and silver.

Work out the probability it contains both gold and silver.

(Total for Question 17 is 4 marks)

18   **(a)** Show that the equation $x^3 - 5x = 9$ has a solution between 2 and 3.

(2)

(b) Show that the equation $x^3 - 5x = 9$ can be rearranged to give $x = \sqrt[3]{9 + 5x}$

(1)

(c) Starting with $x_0 = 3$

Use the iteration formula $x_{n+1} = \sqrt[3]{9 + 5x_n}$ three times to find an estimate for a solution of $x^3 - 5x = 9$

(3)

(Total for Question 18 is 6 marks)

Turn over

19 The measurements of a rectangle are taken.

The length measures 14.4 cm to the nearest mm.

The width measures 32.8 cm to the nearest mm.

By considering bounds work out the area of the rectangle to a suitable degree of accuracy.

You must show all your working and give a reason for your answer.

_____ cm²

(Total for Question 19 is 3 marks)

20 *OBC* is a triangle.

$\overrightarrow{OA} = \mathbf{e} + 2\mathbf{f}$

$\overrightarrow{AD} = \mathbf{e} - \mathbf{f}$

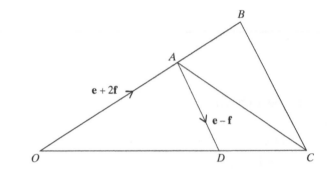

(a) Work out the vector $\overrightarrow{OD}$

_____

(1)

(b) $OA : AB = 2 : 1$ and $\overrightarrow{AC} = 2\mathbf{e} - \dfrac{1}{2}\mathbf{f}$

Prove that *BC* is parallel to *AD*.

(3)

(Total for Question 20 is 4 marks)

21 This solid shape is made from a cylinder and a hemisphere.
The radius of the cylinder and the hemisphere is 8.3 cm.
The height of the shape is 20 cm.

Work out the total surface area of the shape including the base.
Give your answer to 3 significant figures.

| Surface area of a sphere = $4\pi r^2$ |

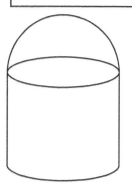

_____ cm²

(Total for Question 21 is 5 marks)

TOTAL FOR PAPER IS 80 MARKS

Answers

Workbook Answers

You are encouraged to show all your working out, as you may be awarded marks for method even if your final answer is wrong. Full marks can be awarded where a correct answer is given without working but, if a question asks for working, you must show it to gain full marks. If you use a correct method that is not shown in the answers below, you would still gain full credit for it.

Page 4: Order and Value

1. a) 0.2491 **[1]**
   b) 53 **[1]**
2. 0.0034 **[1]**
3. 54.8 **[1]**
4. a) 3.5×10^7 **[1]**
   b) 0.003 49 **[1]**
5. a) $6.5 \times 10^2 + 4.2 \times 10^3$ or $650 + 4200$ **[1]**
   = 4850 **[1]**
   b) $(3.1 \times 10^5)^2$ **[1]** = 9.61×10^{10} **[1]**
6. 30×10^6 **[1]** = 3×10^7 **[1]**
7. 0.5×10^4 **[1]** = 5×10^3 **[1]**

Page 5: Types of Number

1. $3^3 \times 5 = 135$ **[1]**
2. $2^3 \times 3^2 \times 5 = 360$ **[1]**
3. $84 = 2^2 \times 3 \times 7$ **[1]**
   $120 = 2^3 \times 3 \times 5$ **[1]**
   HCF = $2^2 \times 3 = 12$ **[1]**
4. 2, 2, 3, 3 **[1]** $2 \times 2 \times 3 \times 3$ or $2^2 \times 3^2$ **[1]**
5. $5^3 = 125$ **[1]**
   125 − 16 = 109 **[1]**
6. Multiples of 12: 12, 24, 36, 48, … **[1]**
   Multiples of 16: 16, 32, 48, … **[1]**
   48 seconds **[1]**
7. $3 \times 6 (= 18)$ **[1]**
8. $4 \times 3 \times 2 \times 1$ **[1]** = 24 **[1]**

Page 6: Basic Algebra

1. $6x + 24 − 15 + 10x$ **[1]**
   = $16x + 9$ **[1]**
2. $2x(4x − 1)$ **[2]**   [1 mark for $2x$ (4x…)]
3. $6x^2 + 2x + 15x − 10$ **[1]**
   $6x^2 + 17x − 10$ **[1]**
4. $2x − x = 1 − 4$ **[1]**
   $x = −3$ **[1]**
5. $5 \times 3 − 2 = 13$ **[1]**
   Perimeter is 4×13 cm **[1]** = 52 cm **[1]**
6. $2x + 35 = 21$ **[1]**
   $2x = 21 − 35$ or $2x = −14$ **[1]**
   $x = −7$ **[1]**
7. $4 + 3x = 2x − 10$ **[1]**
   $3x − 2x = −10 − 4$ **[1]**
   $x = −14$ **[1]**

Page 7: Factorisation and Formulae

1. $3(x − 2)$ **[2]**   [1 mark for a partial factorisation]
2. $8x^2 − 12x + 2x − 3$ **[1]**
   $8x^2 − 10x − 3$ **[1]**
3. $(x + 4)(x − 5)$ **[2]** [1 mark for $(x + a)(x + b)$ where $ab = −20$]
4. $4(x^2 − 25)$ or $(2x − 10)(2x + 10)$ **[1]**
   $4(x − 5)(x + 5)$ **[1]**
5. a) $v^2 − u^2 = 2as$ **[1]**
   $a = \dfrac{v^2 − u^2}{2s}$ **[1]**
   b) $8^2 = 5^2 + 2(a)3$ or $8^2 − 5^2 = 2(a)3$ **[1]**
   $a = 6.5$ **[1]**
6. $r^2 = \dfrac{A}{\pi}$ **[1]**
   $A = \pi r^2$ **[1]**
7. $\dfrac{12(x − 5)}{x(x − 3)(x − 5)} \times \dfrac{x(x − 3)}{3}$ **[3]**
   [1 mark for each correct factorisation]
   $(c =) 4$ **[1]**

Page 8: Ratio and Proportion

1. $\dfrac{6}{11} \times 1650$ or 1 part = £150 **[1]**
   £900 **[1]**
2. $14.4 \div 18$ or $1440 \div 18$ **[1]**
   = 80 cm **[1]**
3. $96 \div 2 = 48$ (men) **[1]**
   $96 − 48 − 12 = 36$ (women) **[1]**
   36 : 12 = 3 : 1 **[1]**
4. $45 \div 3 = 15$ (litres per day) **[1]**
   $15 \times 7 = 105$ litres **[1]**
   Assume same rate **[1]**
5.

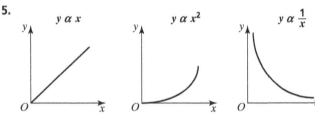

[1 mark for each correct graph]

Page 9: Variation and Compound Measures

1. $210 \div 1.5$ **[1]** = 140 (km/h) **[1]**

   Average speed = Distance ÷ Time

2. 65×3 **[1]** = 195 (miles) **[1]**
3. $4.5 \div 0.3$ or 19.3×0.3 or $4.5 \div 19.3$ **[1]**
   15 or 5.79 or 0.23… and No **[1]**
4. $2000 \times \left(1 + \dfrac{1.5}{100}\right)^2$ or $1500 \times \left(1 + \dfrac{2}{100}\right)^2$ **[1]**

   2060.45 **[1]**
   1560.60 **[1]**
   £60.45 and £60.60 and Money Maker **[1]**

5. $25 \div 0.5$ **[1]** = 50 (mph) **[1]**

Page 10: Angles and Shapes 1 & 2

1. $360 - 64 - 64 = 232$ **[1]**

$232 \div 4$ **[1]** = 58° **[1]**

2. $360 \div 18$ **[1]** = 20 **[1]**

3. $180 - 80 - 62 = 38$ **[1]**

$(180 - 38) \div 2$ **[1]** = 71 **[1]**

$80 - 71 = 9$ **[1]**

4. a) 050° **[1]**

b) $180 + 50$ **[1]** = 230° **[1]**

Page 11: Fractions

1. $\frac{1}{3} = \frac{4}{12}$ and $\frac{5}{6} = \frac{10}{12}$ or $\frac{1}{3} + \frac{5}{6} = \frac{7}{6}$ **[1]**, $\frac{7}{12}$ **[1]**

2. $3 + \frac{3}{6} + 2 + \frac{4}{6}$ or $5 + \frac{7}{6}$ **[1]**, $6\frac{1}{6}$ **[1]**

3. $\frac{6}{20}$ **[1]** = $\frac{3}{10}$ **[1]**

4. $\frac{1}{5}$ is $8 \div 2 = 4$ (people) **[1]**, 4×5 **[1]**, 20 **[1]**

5. $\frac{5}{8} \times 40 = 25$ or $\frac{2}{3} \times 36 = 24$ **[1]**

25 and 24 and $\frac{5}{8}$ of 40 identified **[1]**

6. $\frac{7}{9}$ **[1]**

7. $\frac{1}{3}$ of $180 = 60$ and $\frac{1}{2}$ of $180 = 90$ or $1 - \frac{1}{3} - \frac{1}{2}$ **[1]**

$\frac{30}{180}$ or $\frac{1}{6}$ **[1]**

8. Let $x = 0.3\overset{..}{1}\overset{..}{5}$, $100x = 31.5\overset{..}{1}\overset{..}{5}$ **[1]**

$99x = 31.2$, so $x = \frac{31.2}{99}$ **[1]**

$x = \frac{312}{990}$, so $x = \frac{52}{165}$ **[1]**

Page 12: Percentages

1. $800 \times 1.2 = £960$ **[1]**

$960 \div 12$ **[1]** = £80 **[1]**

2. $\frac{45}{150} \times 100$ (%) **[1]** = 30% **[1]**

3. $£23.50 \times 2 = £47$ **[1]**

$£47 \times 0.9$ **[1]**

$£8.50 \times 5$, £42.30 and £42.50 **[1]** and A cheaper **[1]**.

4. $100\% + 20\% = 120\%$ **[1]**

$120\% \times 0.95 = 114\%$ **[1]**

Increase = 14% **[1]**

5. (£1235 is 95% of sale price)

$1235 \div 0.95$ **[1]** (sale price = £1300)

$£1300 \div 0.8$ **[1]**

Normal price = £1625 **[1]**

Page 13: Probability 1 & 2

1. a)

Letter	A	B	C	D
Probability	0.3	0.1	0.2 **[1]**	0.4 **[1]**

b) 0.3×50 **[1]** = 15 **[1]**

2. a) $(1 - 0.1) \div 3$ or 0.3 **[1]**

$0.3 \times 2 = 0.6$ **[1]**

b) 0.3×120 **[1]** = 36 **[1]**

3. a) $\frac{7}{10}$ **[1]**

b) $\frac{3}{10} \times \frac{3}{10}$ **[1]** = $\frac{9}{100}$ **[1]**

4. a)

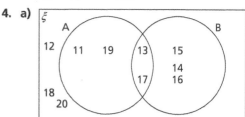

Fully correct **[3]**

[1 mark for 11 and 19 in correct place; 1 mark for 14, 15 and 16 in correct place]

b) $\frac{7}{10}$ **[2]**   [1 mark for identifying correct region]

5. a)

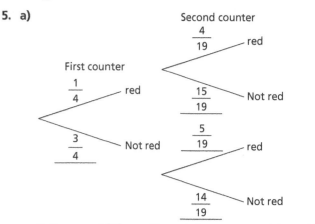

Fully correct **[2]**   [1 mark for any correct branch]

b) $\frac{1}{4} \times \frac{4}{19}$ **[1]** = $\frac{1}{19}$ **[1]**

Page 15: Number Patterns and Sequences & Terms and Rules

1. 12 **[1]**

2. $4n + 6$ **[2]**

[1 mark for $4n + \dots$]

3. $a + b = 1$ and $16a + 4b = 28$ or $4a + b = 7$ **[1]**

$3a = 6$, $a = 2$ **[1]**

$b = -1$ **[1]**

2nd term = 6, 3rd term = 15 **[1]**

4. Linear sequence 17, 14, 11, 8, 5, 2, −1, …. **[1]**

Quadratic sequence 6, 9, 14, 21, …. **[1]**

Quadratic sequence has no negative numbers so only number in both sequences is 14. **[1]**

5. 1st differences are 3   7   11   15

2nd differences are 4   4   4   so coefficient of $n^2 = 4 \div 2 = 2$ **[1]**

If sequence is of the form $2n^2 + bn + c$

1st term: $0 = 2 + b + c$, $-2 = b + c$

2nd term: $3 = 8 + 2b + c$, $-5 = 2b + c$ **[1]**

So $b = -3$ and $c = 1$

nth term is $2n^2 - 3n + 1$ **[1]**

Page 16: Transformations

1. a)
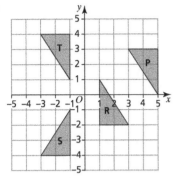

   Fully correct **[2]**
   [1 mark for any 180° rotation of shape **T**]

   b) Reflection **[1]**, in the x-axis (or $y = 0$) **[1]**

   c) $a = 6$ **[1]**, $b = -1$ **[1]**

   > For a translation, sort out the x direction first, then the y direction.

2. a) Fully correct **[2]**
   [1 mark for any reflection in a line $y = c$]

   b) Fully correct **[2]**
   [1 mark for any enlargement scale factor 2]

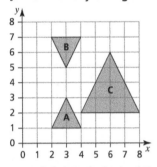

3. Rotation **[1]**; 90° clockwise **[1]**; about (1, 1) **[1]**

Page 17: Constructions

1.
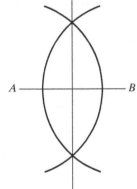

   Fully correct **[2]**
   [1 mark for equal intersecting arcs from A and B]

2.
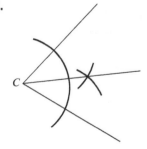

   Fully correct **[2]**
   [1 mark for correct arcs]

3.
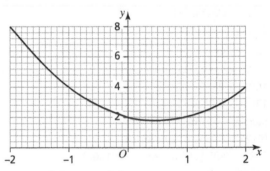

   Fully correct **[2]**
   [1 mark for straight lines 2 cm from given line or semi-circular arcs radius 2 cm]

Page 18: Linear Graphs

1. a)

x	–2	–1	0	1	2
y	–8	–5	–2	1	4

   Fully correct **[2]**
   [1 mark for at least two correct values]

   b)
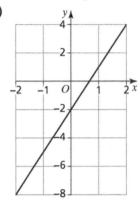

   Fully correct **[2]**
   [1 mark for plotting at least two points correctly]

2. A: $y = 1$ **[1]**; B: $y = x + 1$ **[1]**; C: $y = 1 - x$ **[1]**

3. a) $\dfrac{8 - 2}{3 - 1}$ **[1]** $= 3$ **[1]**

   b) $y = 3x + c$ so $2 = 3(1) + c$ **[1]**
   $c = -1$, $y = 3x - 1$ **[1]**

   c) If C is on the line then when $x = 7$,
   $3 \times 7 - 1 = 20$ so $y = 20$. On the line so
   ABC is a straight line. **[1]**

Page 19: Graphs of Quadratic Functions

1. a)

x	–2	–1	0	1	2
y	8	4	2	2	4

   Fully correct **[2]**
   [1 mark for at least two correct values]

   b)
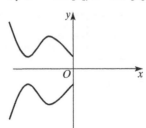

   [2]

   c) $x = -0.6$ **[1]**, $x = 1.6$ **[1]**

2. a) (1, –7) **[1]**
   b) $x = -1.6$ **[1]**, $x = 3.6$ **[1]**

3.

   [1]

Page 20: Powers, Roots and Indices

1. 5 **[1]**

2. 1 or $\frac{1}{49}$ **[1]** $1\frac{1}{49}$ **[1]**

3. a) $64a^3b^6$ **[2]**   [1 mark for two correct terms]
   b) $3x^5y^2$ **[2]**   [1 mark for two correct terms]

4. Not correct, answer is ($d^0 =$) 1 **[1]**

5. $\frac{c^4}{c^3}$ **[1]** $= c$ **[1]**

6. $x = 3$ **[1]**

7. $\frac{1}{8^{\frac{2}{3}}}$ or $\frac{1}{2^2}$ or $\frac{1}{64^{\frac{1}{3}}}$ **[1]** $= \frac{1}{4}$ **[1]**

8. $\frac{10}{\sqrt{5}} = \frac{10 \times \sqrt{5}}{\sqrt{5} \times \sqrt{5}}$ or $\frac{10\sqrt{5}}{5}$ **[1]** $= 2\sqrt{5}$ **[1]**

9. $\frac{\sqrt{3}}{\sqrt{3}-1} = \frac{\sqrt{3}(\sqrt{3}+1)}{(\sqrt{3}-1)(\sqrt{3}+1)}$ **[1]** $= \frac{3+\sqrt{3}}{2}$ **[1]**

Page 21: Area and Volume 1 & 2

1. $\frac{1}{2} \times (10 + 20) \times 8 \ (\times 30)$ **[1]**
   120×30 **[1]** $= 3600$ m³ **[1]**

2. $\pi \times 12^2 \times 15$ or $\frac{1}{2} \times \pi \times 12^2 \times 15$ **[1]**
   6785.8… or 3392.9… **[1]**
   6.7858… or 3.3929… **[1]**
   3.4 litres **[1]**

3. Radius of larger sphere = $4.8 \div 2 \times 3 = 7.2$ **[1]**
   $4 \times \pi \times 7.2^2$ **[1]**
   651.(4…) cm³ **[1]**

4. Volume of cube = $4.5^3 = 91.125$ cm³ **[1]**
   $91.125 = \frac{1}{3} \times 6^2 \times h$ **[1]**
   $91.125 \times 3 \div 36 = 7.59$ cm **[1]**

Page 22: Uses of Graphs

1. $y = 5x - 3$ **[2]**   [1 mark for $y = 5x...$ or gradient = 5]

2. $A(0, 4)$ **[1]** $B(3, 0)$ **[1]**
   Area of triangle = $\frac{1}{2} \times 3 \times 4 = 6$ units² **[1]**

3. e.g. 10 pounds = 4.5 kg **[1]**
   60 pounds is 6×4.5 kg = 27 kg **[1]**

   > Read off at 10 pounds and then scale up.

4. $2y = 3x - 4$ has gradient $\frac{3}{2}$ **[1]**
   Line M has gradient $-\frac{2}{3}$ **[1]**
   Equation of line M is $y = -\frac{2}{3}x + 5$ **[1]**
   $2x + 3y - 15 = 0$ **[1]**

Page 23: Other Graphs 1 & 2

1. a) Area for first 10 seconds = $\frac{1}{2} \times (10 + 20) \times 10$
      = 150 m **[1]**
      Area for last 10 seconds = $\frac{1}{2} \times 10 \times 20 = 100$ m **[1]**
      Total distance is 150 m + 100 m = 250 m **[1]**

   > Distance travelled is the area under the line.

   b) Acceleration is smaller value than deceleration as last line is steeper. **[1]**

   > Acceleration is gradient of the line on a speed–time graph.

2. Draw a tangent at 3 seconds **[1]**
   Make a right-angled triangle using a tangent as shown **[1]**
   Speed = $\frac{9}{2}$ or 4.5 m/s (or correct for suitable tangent) **[1]**

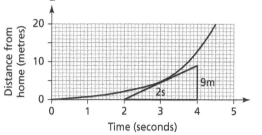

3.
   Fully correct **[2]**
   [1 mark for a circle, with centre at origin, drawn]

4. a) Draw a tangent at 15 seconds **[1]** Make a right angled triangle using the tangent as shown **[1]**
      Gradient = $\frac{21}{19}$ or 1.1 m/s² **[1]**

   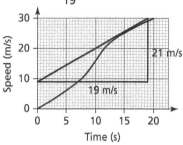

   b) Acceleration **[1]**

   c) Approximate areas =
      $\frac{1}{2} \times 5 \times 6 + \frac{1}{2} \times (6 + 16) \times 5 + \frac{1}{2} \times (16 + 25) \times 5 + \frac{1}{2} \times (25 + 30) \times 5$ or $15 + 55 + 102.5 + 137.5$ **[2]**
      [1 mark for any correct area]
      = 310 m **[1]**

   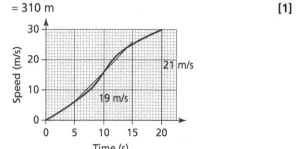

5. $y = \frac{3}{x}$ **[1]** $y = x^3$ **[1]** $y = 3^x$ **[1]**

Page 25: Inequalities

1.   **[2]**
   [1 mark for an open circle at −1 or a filled in circle at 3]

2. $4x - x \geqslant 8 + 7$ or $3x \geqslant 15$ **[1]**
   $x \geqslant 5$ **[1]**

3. $-15 \leqslant 5x \leqslant 10$ **[1]**
   $-3 \leqslant x \leqslant 2$ **[1]**
   9 **[1]**

4. $x + 1 > 3$ or $x + 1 < -3$ **[1]**

$x > 2$ **[1]**

$x < -4$ **[1]**

5. $8n > n^2 + 12$ or $n^2 - 8n + 12 < 0$ **[1]**

$(n - 2)(n - 6) < 0$ **[1]**

$2 < n < 6$ **[1]**

3, 4, 5 **[1]**

6. $24 < x^2 - 1 < 48$ **[1]**

$25 < x^2 < 49$ **[1]**

$5 < x < 7$ **[1]** or $-7 < x < -5$ **[1]**

Page 26: Congruence and Geometrical Problems

1. a) Scale factor 2 or 0.5 or 3.1 cm $\times$ 2 **[1]** = 6.2 cm **[1]**

   b) $5.8 \div 2 = 2.9$ or $5.8 + 3.4 + 6.2 = 15.4$ **[1]**

     $15.4 \div 2$ or $2.9 + 1.7 + 3.1$ **[1]** = 7.7 cm **[1]**

2. $\dfrac{x}{3x + 1} = \dfrac{x + 4}{7x - 2}$ or $x(7x - 2) = (x + 4)(3x + 1)$ **[1]**

$7x^2 - 2x = 3x^2 + x + 12x + 4$ **[1]**

$4x^2 - 15x - 4 = 0$ **[1]**

$(x - 4)(4x + 1) = 0$ **[1]** $x = 4$ **[1]**

> A negative solution is not possible for a length.

3. Angle ABD = Angle BDC (alternate angles) **[1]**

BD is a common side **[1]**

Angle ADB = Angle CBD (alternate angles) **[1]**

ASA so congruent. **[1]**

Page 27: Right-Angled Triangles

1. $13^2 - 4^2$ or $169 - 16$ **[1]**

$\sqrt{169 - 16}$ or $\sqrt{153}$ **[1]** = 12.36... cm **[1]**

2. $7^2 - 3^2$ or $49 - 9$ **[1]**, $\sqrt{49 - 9}$ or $\sqrt{40} = 6.32...$ cm **[1]**

Side of square = 3.32... cm **[1]**

Area = $(3.32...)^2 = 11$ cm^2 **[1]**

3. $\cos PRQ = \dfrac{5.9}{8.6}$ or $\cos PRQ = 0.686...$ **[1]**

= 46.7° **[1]**

4. ACB is bigger as $\sin ACB = 0.375$ and $\sin XZY = 0.333...$

or Angle $ACB = 22°$ and angle $XZY = 19.4...°$ **[1]**

5. $\tan 20° = \dfrac{8.5}{h}$ or $h = \dfrac{8.5}{\tan 20°}$ **[1]**, $h = 23.39...$ **[1]**

Volume = $\dfrac{1}{2} \times 8.5 \times 23.39... \times 24$ **[1]**

= 2382 so 2380 cm^3 **[1]**

Page 28: Sine and Cosine Rules

1.

> Remember the sine rule: $\dfrac{\sin A}{a} = \dfrac{\sin B}{b}$

$\dfrac{\sin A}{9.1} = \dfrac{\sin 28°}{7.8}$ **[1]** $\sin A = \dfrac{9.1 \sin 28°}{7.8}$

or $\sin A = 0.5477...$ **[1]** = 33.2° or 33° **[1]**

2. a)

> Remember the cosine rule: $a^2 = b^2 + c^2 - 2bc \cos A$

$BD^2 = 5.2^2 + 6.9^2 - 2 \times 5.2 \times 6.9 \times \cos 63° = 42.07...$ **[1]**

$BD = \sqrt{5.2^2 + 6.9^2 - 2 \times 5.2 \times 6.9 \times \cos 63°}$ **[1]**

= 6.486... cm **[1]**

   b) $CD = 6.48 \div 5 \times 3 = 3.89...$ **[1]**

$\cos CED = \dfrac{2.1^2 + 4.7^2 - 3.89...^2}{2 \times 2.1 \times 4.7}$ **[1]**

$\cos CED = 0.575...$ **[1]**

Angle $CED = 54.8...°$ or 55° **[1]**

3. $BE^2 = 3.1^2 + 4.6^2$ or $BE = \sqrt{3.1^2 + 4.6^2}$ or 5.54 7... **[1]**

$EC^2 = 3.1^2 + 4.6^2 + 7.8^2$ or

$EC = \sqrt{3.1^2 + 4.6^2 + 7.8^2}$ or 9.57... **[1]**

$\cos BEC = \dfrac{3.1^2 + 4.6^2 + 3.1^2 + 4.6^2 + 7.8^2 - 7.8^2}{2 \times \sqrt{3.1^2 + 4.6^2} \times \sqrt{3.1^2 + 4.6^2 + 7.8^2}}$

or $\cos BEC = \dfrac{5.547^2 + 9.57^2 - 7.8^2}{2 \times 5.547 \times 9.57}$ **[1]**

$\cos BEC = 0.579...$ **[1]**

= 54.58° or 54.6° or 55° **[1]**

Page 29: Statistics 1

1. a) Positive **[1]**

   b) (10, 20) identified **[1]**

   c) Line of best fit drawn **[1]**

     Correct reading from graph, e.g. 12 or 13 **[1]**

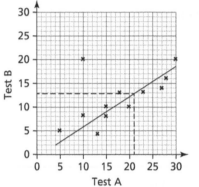

2. a) Position of median is $\dfrac{20 + 1}{2}$ th or 10.5th value, i.e. midway between 10th and 11th value **[1]**

Interval containing median is $4 < t \leqslant 6$ **[1]**

   b) Uses mid-class values 1, 3, 5, 7 **[1]**

$((1 \times 4) + (3 \times 5) + (5 \times 8) + (7 \times 3)) \div 20$ **[1]**

$(4 + 15 + 40 + 21) \div 20 = 4$ minutes **[1]**

3.

Least age	Lower quartile	Median	Upper quartile	Greatest age
17	24	27	37	45

[2]

[1 mark for median; 1 mark for the quartiles]

Page 30: Statistics 2

1. a)

Fully correct **[3]** [1 mark for median in correct place;

1 mark for quartiles in correct place]

   b) 75% (of 60) **[1]** = 45 **[1]**

2. a) First bar represents $5 \times 2 = 10$ buildings or last bar represents $10 \times 1.4 = 14$ buildings **[1]**

Missing bar represents $(10 + 14) \times 2) = 48$ **[1]**

So height of missing bar = (48 ÷ 10) = 4.8 **[1]**

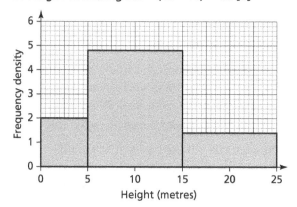

[1]

b) Median height divides the areas in half
$(24 + 48) \div 2 = 36$ **[1]**
So need area of $36 - 10 = 26$ in second bar.
Width to median is $26 \div 4.8 = 5.416...$ or $\frac{26}{48} \times 10$
$= 5.416...$
So estimate of median $= 5 + 5.416... = 10.416...$ m **[1]**

Page 31: Measures, Accuracy and Finance

1. 17.55 **[1]**, 17.65 **[1]**
2. a) 150×1.38 **[1]** $207 **[1]** $70 + \$125 = \$195$ and
   enough **[1]**
   b) $70 \div 1.38$ or 59×1.38 **[1]**
   £50.72 or $81.42 and cheaper in USA. **[1]**
3. a) $\frac{200 \times 10}{50}$ **[2]** [1 mark for two correct approximations]
   $= 40$ **[1]**
   b) Overestimate as numerator increased and
   denominator decreased **[1]**
4. Stating bound of 5.355 or 5.365 **[1]**
   Using $A = \pi r^2$ to obtain lower bound for
   $A = 90.088...$ or upper bound for $A = 90.425...$ **[1]**
   90.088 and 90.425 **[1]**
   90 with supporting reason, e.g. both bounds
   round to 90 **[1]**

Page 32: Solving Non-Linear Equations

1. $(x - 5)(x + 3)$ **[1]**, $x = 5$ or $x = -3$ **[1]**
2. a) $x(x + 5) = 84$ **[1]**
   b) $x^2 + 5x = 84$ or $x^2 + 5x - 84 = 0$ **[1]**
   $(x - 7)(x + 12)(= 0)$ **[1]**
   $x = 7$ **[1]**
3. a) $a = 4$ **[1]**
   $(x - 4)^2 = x^2 - 8x + 16$, so $b = -14$ **[1]**
   b) $(4, -14)$ **[1]**
4. a) When $x = 2$, $x^3 + 2x = 8 + 4 = 12$ **[1]**
   When $x = 3$, $x^3 + 2x = 27 + 6 = 33$, so solution
   between 2 and 3 **[1]**
   Alternative method:
   When $x = 2$, $x^3 + 2x - 20 = 8 + 4 - 20 = -8$ **[1]**
   When $x = 3$, $x^3 + 2x - 20 = 27 + 6 - 20 = 13$, so
   solution between 2 and 3 **[1]**
   b) $x^3 = 20 - 2x$, so $x = \sqrt[3]{20 - 2x}$ **[1]**

c) Using a calculator, input 2 and press "=", then input
$\sqrt[3]{20 - 2(\text{Ans})}$ and pressing "=" repeatedly gives
2, 2.5198... **[1]**, 2.4640... **[1]**, 2.4701..., 2.4694...,
2.4695..., 2.4695...
So estimate of solution is 2.4695 or 2.47 **[1]**

Page 33: Simultaneous Equations and Functions

1. $4x + 2y = 2$ or $2x - 4y = 16$ **[1]**
   $5x = 10$ or $5y = -15$ **[1]**
   $x = 2$ or $y = -3$ **[1]**
   $x = 2$ and $y = -3$ **[1]**
2. $y = \frac{2 - 3x}{2}$ or $x = \frac{2 - 2y}{3}$ **[1]**
   $5x^2 - \left(\frac{2 - 3x}{2}\right)^2 = 4$ or $5\left(\frac{2 - 2y}{3}\right)^2 - y^2 = 4$ **[1]**
   $20x^2 - (4 - 12x + 9x^2) = 16$ or $11x^2 + 12x - 20 = 0$
   or $5(4 - 8y + 4y^2) - 9y^2 = 36$ or $11y^2 - 40y - 16 = 0$ **[1]**
   $(x + 2)(11x - 10) = 0$ or $(y - 4)(11y + 4) = 0$ **[1]**
   $x = -2$, $y = 4$ or $x = \frac{10}{11}$, $y = -\frac{4}{11}$ **[1]**
3. a) $f(-4) = (-4)^2 + 3 = 19$ **[1]**
   b) $fg(5) = f[g(5)] = f(20 - 1) = f(19)$ **[1]** $= 19^2 + 3 = 364$ **[1]**
4. a) Let $y = 5x^2 + 1$ Rearranging gives $\sqrt{\frac{y - 1}{5}} = x$
   so $f^{-1}(x) = \sqrt{\frac{x - 1}{5}}$ **[1]**
   $f^{-1}(6) = \sqrt{\frac{6 - 1}{5}} = \sqrt{\frac{5}{5}} = 1$ as $x = 0$ **[1]**
   b) $gf(x) = g[f(x)] = g(5x^2 + 1) = 2(5x^2 + 1)$ **[1]**
   $2(5x^2 + 1) = 3x + 3$ so $10x^2 - 3x - 1 = 0$ **[1]**
   $(5x + 1)(2x - 1) = 0$ **[1]**
   $x = -\frac{1}{5}$ or $x = \frac{1}{2}$ **[1]**

Page 34: Algebraic Proof

1. Let 1st odd number be $2n + 1$, so next odd number is
   $2n + 3$ **[1]**
   $(2n + 1)^2 + (2n + 3)^2 = 4n^2 + 4n + 1 + 4n^2 + 12n + 9$ **[1]**
   $= 8n^2 + 16n + 10 = 2(4n^2 + 8n + 5)$ so multiple of 2 **[1]**
2. Let 1st even number be $2n$, so next even number is $2n + 2$ **[1]**
   $(2n + 2)^2 - (2n)^2 = 4n^2 + 8n + 4 - 4n^2$ **[1]**
   $= 8n + 4 = 4(2n + 1)$ so multiple of 4 **[1]**
3. Let the consecutive numbers be $n, n + 1, n + 2, n + 3, n + 4$
   So sum $= n + n + 1 + n + 2 + n + 3 + n + 4 = 5n + 10$ **[1]**
   $= 5(n + 2)$, so multiple of 5 **[1]**
4. $\frac{1}{x + 1} - \frac{4}{x^2 + 6x + 5} = \frac{1}{x + 1} - \frac{4}{(x + 1)(x + 5)}$ **[1]**
   $= \frac{x + 5}{(x + 1)(x + 5)} - \frac{4}{(x + 1)(x + 5)}$ **[1]**
   $= \frac{(x + 1)}{(x + 1)(x + 5)}$ **[1]**
   $= \frac{\cancel{(x + 1)}}{\cancel{(x + 1)}(x + 5)} = \frac{1}{x + 5}$ **[1]**
5. a) $\frac{5}{x - 1} + \frac{2}{x + 1} = 1$ so multiply through by common
   denominator
   $5(x + 1) + 2(x - 1) = (x + 1)(x - 1)$ **[2]** [1 mark for each side]
   $5x + 5 + 2x - 2 = x^2 - 1$ **[1]**
   $0 = x^2 - 7x - 4$ **[1]**

b)

> Remember the quadratic formula:
> $$x = \frac{-b \pm \sqrt{b^2 - 4ac}}{2a}$$

$a = 1, b = -7, c = -4$ $\quad x = \dfrac{-(-7) \pm \sqrt{(-7)^2 - 4(1)(-4)}}{2(1)}$ **[1]**

$x = \dfrac{7 \pm \sqrt{65}}{2}$ **[1]**

Page 35: Circles

1. Angle $BDC = 90°$
   (Angle in a semicircle, angle on a diameter) **[1]**
   Angle $DBC = 180° - 64° - 90° = 26°$
   (Angles in a triangle) **[1]**
   Angle $ADB = 26°$ (Alternate angles) **[1]**
2. Angle $RTO = 90°$ (Tangent to radius = 90°) **[1]**
   Angle $ROT = 180° - 37° - 90° = 53°$
   (Angles in a triangle) **[1]**
   Angle $POT = 180° - 53° = 127°$ (Angles on a straight line)
   So angle $OTP = 53° \div 2 = 26.5°$ (Isosceles triangle) **[1]**
   Angle $PTS = 90° - 26.5° = 63.5°$ **[1]**
3. Angle $BCD = x$ (Angle between tangent and chord = angle at circumference subtended by the chord) **[1]**
   Angle $CBD = x$ (Isosceles triangle) so
   angle $CDB = 180° - 2x$ (Angles in triangle) **[1]**
   Angle $BAC = 180° - (180° - 2x) = 2x$ (Opposite angles of a cyclic quadrilateral add up to 180°) **[1]**
4. Angle $OBC =$ angle $OCB = 15°$ (isosceles triangle) **[1]**
   Angle $BOC = 150°$ (angles in a triangle sum to 180°) **[1]**
   Angle $CAB = \dfrac{150}{2} = 75°$ (angle at circumference
   $= \dfrac{1}{2}$ angle at centre) **[1]**

Page 36: Vectors

1. $\begin{pmatrix} 18 \\ -5 \end{pmatrix}$ **[2]** [1 mark for each value]

2. a) $\overrightarrow{OC} = \mathbf{a} + \mathbf{b}$ **[1]**
   $\overrightarrow{OP} = \dfrac{1}{3}(\mathbf{a} + \mathbf{b})$ or $\overrightarrow{OP} = \dfrac{1}{3}\mathbf{a} + \dfrac{1}{3}\mathbf{b}$ **[1]**
   b) $\overrightarrow{AP} = \overrightarrow{AO} + \overrightarrow{OP} = -\mathbf{a} + \dfrac{1}{3}\mathbf{a} + \dfrac{1}{3}\mathbf{b}$ **[1]**
   $\overrightarrow{AP} = -\dfrac{2}{3}\mathbf{a} + \dfrac{1}{3}\mathbf{b}$ **[1]**
   c) $\overrightarrow{PM} = \overrightarrow{PO} + \overrightarrow{OM} = -\dfrac{1}{3}\mathbf{a} - \dfrac{1}{3}\mathbf{b} + \dfrac{1}{2}\mathbf{b}$ **[1]**
   $\overrightarrow{PM} = -\dfrac{1}{3}\mathbf{a} + \dfrac{1}{6}\mathbf{b}$ **[1]**
   $AP : PM = 2 : 1$ **[1]**
3. a) $\overrightarrow{TA} = \dfrac{4}{5}\mathbf{a}$ **[1]**
   b) $\overrightarrow{AB} = \overrightarrow{AO} + \overrightarrow{OB} = -\mathbf{a} + \mathbf{b}$ **[1]**
   $\overrightarrow{AS} = \dfrac{5}{7}(-\mathbf{a} + \mathbf{b})$ **[1]**
   $\overrightarrow{TS} = \overrightarrow{TA} + \overrightarrow{AS} = \dfrac{4}{5}\mathbf{a} + \dfrac{5}{7}(-\mathbf{a} + \mathbf{b})$ **[1]**
   $\overrightarrow{TS} = \dfrac{3}{35}\mathbf{a} + \dfrac{5}{7}\mathbf{b}$ **[1]**

Pages 37–50: Practice Exam Paper 1

1. a) $162 - 50 = 112$ miles at 70 mph **[1]**
   $112 \div 70$ or $11.2 \div 7$ or $50 \div 50$ or 1 **[1]**
   1.6 **[1]**
   $1.6 + 1 = 2.6$ hours or 2 hours 36 minutes **[1]**
   b) Average speed was less than assumed. **[1]**
2. $3 \times (-4)^2 \times 2$ or $3 \times -4 \times -4 \times 2$ **[1]** $= 96$ **[1]**
3. $\dfrac{140 \times 2}{70}$ **[2]** [1 mark for two of 140 or 100, 2 and 70]
   Answer 4 (if 140 used) or 3 (if 100 used) **[1]**
4. Each angle at centre $= 360° \div 5 = 72°$ **[1]**
   In a triangle this leaves $180° - 72° = 108°$ **[1]**
   Triangle is isosceles so $108° \div 2 = 54°$ **[1]**
5. Box A $P(\text{blue}) = \dfrac{4}{10}$ or $\dfrac{2}{5}$ or 0.4 **[1]**
   Box B $P(\text{blue}) = \dfrac{6}{16}$ or $\dfrac{3}{8}$ or 0.375 **[1]**
   Not correct as $0.4 > 0.375$ **[1]**
6. a) $4x - 20 = 24$ **[1]**
   $4x = 24 + 20$ or $4x = 44$ **[1]**
   $x = 11$ **[1]**
   b) $4x(2x - 3)$ **[2]**
   [1 mark for partial factorisation, e.g. $2x(4x - 6)$
   or $x(8x - 12)$]
7. a) 2.37×10^5 **[1]**
   b) 0.000 45 **[1]**
   c) 5.62×10^4 is greater as a larger power of 10 **[1]**
8. Assume, for example, 100 students in the school
   so 60 boys and 40 girls **[1]**
   $60 \div 4 = 15$ boys study French and 20 girls study French **[1]**
   $15 + 20 = 35\%$ study French **[1]**
9. $(x - 4)(x - 7)$ **[2]** [1 mark for $(x - a)(x - b)$ where $ab = 28$]
   $x = 4$ or $x = 7$ **[1]**

> You need to factorise into two brackets first.

10. $\begin{pmatrix} 4 \\ 9 \end{pmatrix} + 2\begin{pmatrix} 3 \\ -2 \end{pmatrix} = \begin{pmatrix} 4 \\ 9 \end{pmatrix} + \begin{pmatrix} 6 \\ -4 \end{pmatrix} = \begin{pmatrix} 10 \\ 5 \end{pmatrix}$ **[2]**

[1 mark for each part]
11. Angle $DBC = 66°$ (Isosceles triangle) **[1]**
    Angle $BDC = 180° - 66° - 66° = 48°$
    (Angles in a triangle add up to 180°) **[1]**
    Angle $BAT = 32°$ or Angle $CBS = 48°$
    (Angle between tangent and chord is equal to angle subtended on the chord) **[1]**
    Angle $ABC = 180° - 32° - 48° = 100°$
    (Angles on a straight line add up to 180°) **[1]**
12. a)

Age (A years)	Frequency	Age (A years)	Cumulative frequency
$0 < A \leqslant 20$	15	$\leqslant 20$	15
$20 < A \leqslant 40$	36	$\leqslant 40$	51
$40 < A \leqslant 60$	32	$\leqslant 60$	83
$60 < A \leqslant 80$	17	$\leqslant 80$	100

[1]

Fully correct graph plotted at upper class boundaries **[1]**

b) Reading from graph at 75 **[1]**; 54 years **[1]**

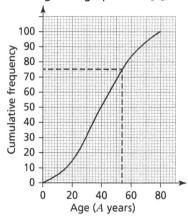

13. a) $\sqrt{18} - \sqrt{6} + \sqrt{6} - \sqrt{2}$ **[1]**
$\sqrt{2}(\sqrt{9} - 1)$ **[1]**
$2\sqrt{2}$, $a = 2$ **[1]**

b) $\dfrac{10(3 - \sqrt{5})}{(3 + \sqrt{5})(3 - \sqrt{5})}$ **[1]**

$\dfrac{10(3 - \sqrt{5})}{(9 - 5)} = \dfrac{10(3 - \sqrt{5})}{4}$ **[1]**

$\dfrac{5(3 - \sqrt{5})}{2}$ **[1]**

14. Let $x = 0.1\dot{7}$, $10x = 1.\dot{7}$ **[1]**
$9x = 1.6$, $x = \dfrac{1.6}{9}$ **[1]**
$x = \dfrac{16}{90}$, $x = \dfrac{8}{45}$ **[1]**

15. a) $\left(\dfrac{2}{5}\right)^2$ **[1]** $\dfrac{4}{25}$ **[1]**

> Always work out the root before the power.

b) $3^{-4} = \dfrac{1}{81}$ so $x = -4$ **[1]**

16. a) $y = kx^2$ **[1]** $x = 4$, $y = 48$, so $48 = k(4)^2$ **[1]**
$k = 3$, $y = 3x^2$ **[1]**

b) $27 = 3x^2$ or $9 = x^2$ **[1]** $x = 3$ **[1]**

17. Volume of hemisphere $= \dfrac{2}{3} \times \pi \times 3^3$ or 18π cm^3 **[1]**

Volume of cone $= \dfrac{2}{3} \times \pi \times 3^3 \div 3 \times 8$ or $18\pi \div 3 \times 8$ or 48π cm^3 **[1]**
$48\pi = \dfrac{1}{3} \times \pi \times 6^2 \times h$ or $48\pi = 12\pi h$ **[1]**
$h = 4$ **[1]**

18. $\tan 30° = \dfrac{1}{\sqrt{3}}$ and $\cos 60° = \dfrac{1}{2}$ **[1]**

$\dfrac{1}{(\tan 30°)^2} \times \dfrac{1}{\cos 60°} = \dfrac{1}{\left(\dfrac{1}{\sqrt{3}}\right)^2} \times \dfrac{1}{\left(\dfrac{1}{2}\right)}$ or $3 \times 2 = 6$ **[1]**

19. Let number of counters be yellow $= 3x$ and red $= 4x$

Probability first counter is yellow $= \dfrac{3}{7}$ **[1]**
After 6 yellow counters are added, yellow $= 3x + 6$
and red is still $4x$ **[1]**
Probability both yellow $= \dfrac{3}{7} \times$ probability second is
yellow $= \dfrac{3}{14}$
or probability second is yellow $= \dfrac{3}{14} \div \dfrac{3}{7} = \dfrac{1}{2}$ **[1]**
So $\dfrac{3x + 6}{7x + 6} = \dfrac{1}{2}$ or $6x + 12 = 7x + 6$ **[1]**
$x = 6$ and number of red counters $= 4 \times 6 = 24$ **[1]**

20. $3x^2 = 2(4 - x)$ or $3x^2 = 8 - 2x$ **[1]**
$3x^2 + 2x - 8 = 0$ **[1]**
$(3x - 4)(x + 2)(= 0)$ **[1]**
$x = \dfrac{4}{3}$ or $x = -2$ **[1]**

21. a)

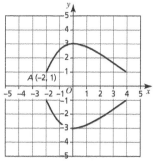

[1]

b) $(-3, -1)$ **[1]**
c) $(0, 3)$ **[1]**

22. Parabola drawn with intercept at $(0, -1)$ **[1]**
Use a method to find the roots of $y = 0$

E.g. $x = \dfrac{-(-8) \pm \sqrt{(-8)^2 - 4(2)(1)}}{2(2)}$ or $2(x - 2)^2 - 9 (= 0)$ **[1]**

Roots found, e.g. $2 \pm \sqrt{\dfrac{9}{2}}$ **[1]**

Turning point is $(2, -9)$ **[1]**
Fully correct parabola drawn with Turning point at
$(2, -9)$ and intercepts at $(0, -1)$, $(2 + \sqrt{\dfrac{9}{2}})$ $(2 - \sqrt{\dfrac{9}{2}})$
clearly shown **[1]**

Pages 51–64: Practice Exam Paper 2

1. $(7 + x + 3x + 6 + 9) \div 5 = 40$
$7 + x + 3x + 6 + 9 = 40 \times 5$ **[1]**
$4x + 22 = 200$ or $4x = 178$ **[1]** $x = 44.5$ **[1]**

2. a) Mass is 2.5×9 **[1]** $= 22.5$ g **[1]**

> Mass = Volume × Density

b) Pressure decreases **[1]**

3. a) Factors of 36: 1, 2, 3, 4, 6, 9, 12, 18, 36 or $36 = 2^2 \times 3^2$
Factors of 54: 1, 2, 3, 6, 9, 18, 27, 54 or $54 = 2 \times 3^3$ **[1]**
HCF $= 2 \times 3^2 = 18$ **[1]**

b) Multiples of 9: 9, 18, 27, 36, … or $9 = 3^2$
Multiples of 12: 12, 24, 36, 48, … or $12 = 2^2 \times 3$
Multiples of 18: 18, 36, 54, 72, … or $18 = 2 \times 3^2$ **[1]**
LCM $= 2^2 \times 3^2 = 36$ **[1]**

4. Rotation **[1]** 90° clockwise **[1]** about $(3, 1)$ **[1]**

5. a) 6303.878201 **[2]** [1 mark for $\dfrac{62086}{9.8488…}$]
b) 6300 **[1]**

6. $4 : 10 = y : 15$ or $\dfrac{4}{10} = \dfrac{y}{15}$ **[1]**
$y = 6$ **[1]**

7. a) Not fair as frequencies vary too much **[1]**
b) $9 + 11 = 20$ **[1]**
$\dfrac{20}{60}$ or $\dfrac{1}{3}$ **[1]**

8. a) 15 out of 50 or $\dfrac{15}{50}$ is 4500 customers so 30% or
$4500 \div 30 \times 100$ **[1]** $= 15\,000$ **[1]**

b) Assume the sample is representative of all
customers **[1]**

9. $11.25 \div 0.75 = 15$ books (smaller pile) **[1]**
$15 \div 3 \times 5$ **[1]** $= 25$ books (larger pile) **[1]**
40 books altogether **[1]**

10. $1508.5 \text{ m} \leqslant L < 1509.5 \text{ m}$ **[2]** [1 mark for one correct]

11. a) $f(-5) = 3(-5)^2 = 3 \times 25 = 75$ **[1]**

   b) $fg(10) = f[g(10)] = f\left(\dfrac{1}{2}\right)$ **[1]**

   $= 3 \times \left(\dfrac{1}{2}\right)^2 = \dfrac{3}{4}$ **[1]**

   c) $h^{-1}(x) = \sin^{-1}\left(\dfrac{x}{2}\right)$ **[1]**

   $h^{-1}(0.4) = \sin^{-1}\left(\dfrac{0.4}{2}\right) = \sin^{-1}(0.2)$ **[1]**

   $= 11.5°$ or $12°$ **[1]**

   > Work out $h^{-1}(x)$ first.

12. $(4x + 3)(2x + 1)(x - 1) = (8x^2 + 4x + 6x + 3)(x - 1)$ or

   $(8x^2 + 10x + 3)(x - 1)$ **[1]**

   $= 8x^3 - 8x^2 + 10x^2 - 10x + 3x - 3$ **[1]**

   $= 8x^3 + 2x^2 - 7x - 3$ **[1]**

13. Area of small circle $= \pi \times 1^2$ or

   Area of large circle $= \pi \times 4.5^2$ **[1]**

   Area of sector $= \dfrac{360° - 25°}{360°} \times \pi \times 4.5^2$ **[1]**

   Required area $= \dfrac{360° - 25°}{360°} \times \pi \times 4.5^2 - \pi \times 1^2$ **[1]**

   $= 59.199... - 3.1415...$ **[1]**

   $= 56.05... = 56.1 \text{ cm}^2$ **[1]**

14. a) A tangent drawn at 10 seconds **[1]**

   Gradient $= \dfrac{17}{12.6}$ **[1]** (readings may vary depending on triangle drawn)

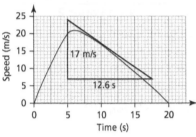

   Gradient $= -1.35$ **[1]** (Accept -1.3 to -1.4)

   b) Gradient represents acceleration. **[1]**

   > Note that the answer to part a) is negative as the car is decelerating.

   c)
   > Distance = area under curve

   Area of triangle $= \dfrac{1}{2} \times 5 \times 20 = 50 \text{ m}$ **[1]**

   Area of trapezium $= \dfrac{1}{2} \times (20 + 17) \times 5 = 92.5 \text{ m}$ **[1]**

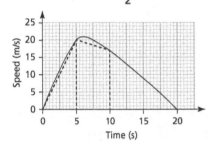

   Total distance $= 50 + 92.5 = 142.5 \text{ m}$ **[1]**

   > Distance travelled is worked out using the area under the curve on a speed–time graph.

15. Let the two odd numbers be $2x + 1$ and $2y + 1$ **[1]**

   $(2x + 1)^2 + (2y + 1)^2 = 4x^2 + 4x + 1 + 4y^2 + 4y + 1$ **[1]**

   $= 2(2x^2 + 2y^2 + 2x + 2y + 1)$ therefore even **[1]**

16.
   > Sine rule is $\dfrac{a}{\sin A} = \dfrac{b}{\sin B} = \dfrac{c}{\sin C}$

   $\dfrac{AD}{\sin 43°} = \dfrac{8.2}{\sin 105°}$ **[1]**

   $AD = \dfrac{8.2 \sin 43°}{\sin 105°}$ or $AD = 5.789...$ cm **[1]**

   $BD = 8.2 \div 2 \times 3$ or $BD = 12.3$ cm **[1]**

   > Cosine rule is $a^2 = b^2 + c^2 - 2bc \cos A$

   $AB^2 = (5.789...)^2 + (12.3)^2 - 2 \times 5.789... \times 12.3 \times \cos 64°$ **[1]**

   $AB =$
   $\sqrt{(5.789...)^2 + (12.3)^2 - 2 \times 5.789... \times 12.3 \times \cos 64°}$ **[1]**

   $AB = 11.06232... = 11.1$ cm **[1]**

17. Each team plays 23 matches so number of matches $=$

   $\dfrac{24 \times 23}{2}$ (24×23 would count each match twice; once for each team) **[1]** $= 276$ **[1]**

18. $\dfrac{x - 4}{y - 4} = \dfrac{1}{4}$ or $\dfrac{x + 6}{y + 6} = \dfrac{1}{2}$ **[1]**

   $4x - 16 = y - 4$ or $2x + 12 = y + 6$

   or $4x - 12 = y$ and $2x + 6 = y$ **[1]**

   $4x - 12 = 2x + 6$ or $2x = 18$ or $x = 9$ **[1]**

   $y = 2(9) + 6$ or $y = 24$ **[1]**

   $9 : 24 = 3 : 8$ **[1]**

19. a) As ratio is $2 : 3 : 4$ number of shapes must be a multiple of $2 + 3 + 4 = 9$ **[1]**

   As labels are in ratio $1 : 6$ must also be a multiple of 7 so must be a multiple of $9 \times 7 = 63$ so least number in bag is 63 which is greater than 60. **[1]**

   b) If 63 shapes in bag then number of circles

   $= \dfrac{2}{9} \times 63$ **[1]** $= 14$ **[1]**

   So number of circles labelled A $= \dfrac{1}{7} \times 14 = 2$ **[1]**

   So probability chosen $= \dfrac{2}{63}$ **[1]**

   Alternative method:

   Probability a circle $= \dfrac{2}{9}$ **[1]**

   Probability labelled A $= \dfrac{1}{7}$ **[1]**

   Probability a circle and labelled A $= \dfrac{2}{9} \times \dfrac{1}{7}$ **[1]** $= \dfrac{2}{63}$ **[1]**

20.
   > Area of a triangle $= \dfrac{1}{2} ab \sin C$

   Area $= \dfrac{1}{2} \times 15.8 \times 12.4 \times \sin 72°$ **[1]**

   $= 93.16...$ cm^2 or 93.2 cm^2 **[1]**

21. a)
   > Frequency density = Frequency ÷ Class width

Time (t minutes)	Frequency	Class width (minutes)	Frequency density
$0 < t \leqslant 1$	8	1	8
$1 < t \leqslant 4$	9	3	3
$4 < t \leqslant 6$	18	2	9
$6 < t \leqslant 10$	25	4	6.25

   [1] (8, 3, 9 and 6.25 seen)

   Label vertical axis "frequency density" **[1]**

Fully correct graph **[1]**

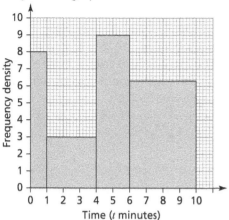

b) Median divides area in half so at $60 \div 2 = 30$

First two bars represent 8 and 9 so $30 - 8 - 9 = 13$ **[1]**

Median $= 4 + \frac{13}{18} \times 2 = 5.4\ldots$ minutes or $5\frac{4}{9}$ minutes **[1]**

Pages 65–77: Practice Exam Paper 3

1. $3.6^2 + BC^2 = 9.4^2$ or $BC^2 = 9.4^2 - 3.6^2$ **[1]**

$BC^2 = \sqrt{9.4^2 - 3.6^2} = 8.68\ldots$ **[1]**

8.7 cm **[1]**

2. 5^3 **[2]** [1 mark for each part of $\frac{5^{10}}{5^7}$] $= 125$ **[1]**

3. P(B wins) $= 0.54 \div 3 \times 2$ **[1]** $= 0.36$ **[1]**

P(A wins) $= 1 - 0.54 - 0.36$ **[1]** $= 0.1$ **[1]**

4. **a)** $5 < t \leqslant 10$ **[1]**

b)

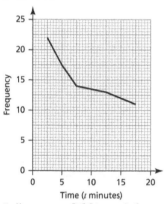

Fully correct **[2]** [1 mark for correct points used]

> Plot the points for a frequency polygon at mid-class values.

5. Number of litres $= 250 \times 4.5$ **[1]** $= 1125$ **[1]**

$1125 \div 12 = 93.75$ minutes **[1]**

1 hour 34 minutes **[1]**

6. **a)**

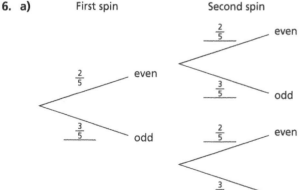

Fully correct **[2]** [1 mark for one correct odd branch]

b) $\frac{2}{5} \times \frac{2}{5}$ **[1]** $= \frac{4}{25}$ **[1]**

7. $6x + 4y = 22$ or $3x - 12y = 39$ **[1]**

$7x = 35$ or $14y = -28$ or $x = 5$ or $y = -2$ **[1]**

$x = 5$ and $y = -2$ **[1]**

> Either multiply the first equation by 2 on both sides or multiply the second equation by 3 on both sides to match coefficients.

8. **a)**

x	−1	0	1	2	3	4	5	6
y	8	2	−2	−4	−4	−2	2	8

Fully correct **[2]** [1 mark for at least three correct values]

b)

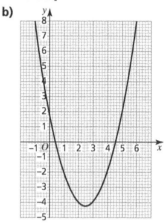

[2]

c) $(2.5, -4.25)$ **[2]** (accept −4.2 or −4.3 for y coordinate; 1 mark for each coordinate)

9. Angle in sector $= 180° - 68° = 112°$ **[1]**

Radius of sector $= 18$ m or Circumference of circle $= 2 \times \pi \times 18$ **[1]**

Arc length $= \frac{112}{360} \times 2 \times \pi \times 18$ or $35.185 \ldots$ m **[1]**

Perimeter $= \frac{112}{360} \times 2 \times \pi \times 18 + 23 + 23 + 18 + 18$ **[1]**

Perimeter $= 117.185\ldots = 117$ m **[1]**

10. $2(2)^2 + c = 9$ or $c = 9 - 2(2)^2$ $c = 1$ **[1]**

7th term $= 2(7)^2 + 1$ **[1]** $= 99$ **[1]**

11. Multiplier for first 2 years is 1.015 or multiplier for last 3 years is 1.0075 **[1]**

After five years value of account $=$
£2780 $\times 1.015^2 \times 1.0075^3$ **[1]**

$=$ £2928.95 so not correct. **[1]**

12. Probability of an even number $= \sqrt[3]{\frac{27}{125}} = \frac{3}{5}$ **[1]**

So probability of getting two odd numbers
$= \left(\frac{2}{5}\right)^2 = \frac{4}{25}$ **[1]**

13. Speed $= 9.6 \times 1.236 \times 10^3 = 11865.6$ km/h **[1]**

$= 11865.6 \div 60$ km/min **[1]** $= 197.76$ km (in 1 minute) **[1]**

14. $c(d + 1) = 5d - 2$ or $cd + c = 5d - 2$ **[1]**

$c + 2 = 5d - cd$ or $cd - 5d = -2 - c$ **[1]**

$c + 2 = d(5 - c)$ or $d(c - 5) = -2 - c$ **[1]**

$d = \frac{-2 - c}{c - 5}$ or $d = \frac{c + 2}{5 - c}$ **[1]**

15. Let height of new tetrahedron be H.

Volume of original tetrahedron $V = \frac{1}{3} \times A \times h$ so for

new tetrahedron $V = \frac{1}{3} \times 1.25A \times H$ **[1]**

$\frac{1}{3} \times A \times h = \frac{1}{3} \times 1.25A \times H$ or $h = 1.25 \times H$ [1]

$H = \frac{h}{1.25} = \frac{4h}{5}$ or $H = 80\%$ of h so percentage decrease = 20%   [1]

16. Gradient of line $L = -\frac{5}{2}$ [1]

    Gradient of perpendicular line $= \frac{2}{5}$ [1]

    Coordinates of A (2, 0) and coordinates of B (0, 5)   [1]

    Coordinates of midpoint of AB $(1, \frac{5}{2})$ [1]

    Equation of perpendicular bisector is $y - \frac{5}{2} = \frac{2}{5}(x - 1)$
so $10y - 4x - 21 = 0$ [1]

17. $x : y : z = 18 : 3 : 10$ **[1]** so $18 + 3 + 10 = 31$ parts
$= 80 - 18 = 62$ **[1]**

    1 part = 2, so $x = 36$, $y = 6$ and $z = 20$ or
3 parts out of 31 parts   [1]

    Required probability $= \frac{6}{62}$ or $\frac{3}{31}$ [1]

18. a) When $x = 2$, $x^3 - 5x = 2^3 - 5(2) = 8 - 10 = -2$ [1]
      When $x = 3$, $x^3 - 5x = 3^3 - 5(3) = 27 - 15 = 12$
so solution between 2 and 3   [1]

      Alternative method: When $x = 2$, $x^3 - 5x - 9 =$
$2^3 - 5(2) - 9 = 8 - 10 - 9 = -11$ [1]
      When $x = 3$, $x^3 - 5x - 9 = 3^3 - 5(3) - 9 =$
$27 - 15 - 9 = 3$, so solution between 2 and 3   [1]

    b) $x^3 = 5x + 9$ and $x = \sqrt[3]{9 + 5x}$ [1]

    c) Using a calculator, input 3 and press "=", then input
$\sqrt[3]{9 + 5(\text{Ans})}$ and pressing "=" repeatedly gives 3,

2.8844... **[1]**, 2.8611... **[1]**, 2.8564..., so estimate of solution is 2.86 **[1]**

19. Bounds of 14.35, 14.45, 32.75 and 32.85   [1]
    Area for lower bounds = $14.35 \times 32.75 = 469.96...$ cm^2
    (or Area for upper bounds = 14.45×32.85) = 474.68... **[1]**
    469.96... and 474.68... and chooses 470 cm^2 as
both answers round to 470   [1]

20. a) $\overrightarrow{OD} = \overrightarrow{OA} + \overrightarrow{AD} = \mathbf{e} + 2\mathbf{f} + \mathbf{e} - \mathbf{f} = 2\mathbf{e} + \mathbf{f}$ [1]
    b) $\overrightarrow{AB} = \overrightarrow{OA} = \frac{1}{2}\mathbf{e} + \mathbf{f}$ [1]

      $\overrightarrow{BC} = \overrightarrow{BA} + \overrightarrow{AC} = -\frac{1}{2}\mathbf{e} - \mathbf{f} + 2\mathbf{e} - \frac{1}{2}\mathbf{f} = \frac{3}{2}\mathbf{e} - \frac{3}{2}\mathbf{f}$ [1]
      BC is $\frac{3}{2} \times AD$ so as BC is a multiple of AD, they are parallel.   [1]

21. Base area $= \pi \times 8.3^2$ or 216.42...
    Curved surface area of cylinder = $2 \times \pi \times 8.3 \times (20 - 8.3)$
or 610.16...
    Curved surface area of hemisphere $= 2 \times \pi \times 8.3^2$ or
432.84 ...
    [3] [1 mark for each correct area; base and hemisphere
may be combined for 2 marks]
    $\pi \times 8.3^2 + 2 \times \pi \times 8.3 \times (20 - 8.3) + 2 \times \pi \times 8.3^2$ or
    216.42... + 610.16... + 432.84 ...   [1]
    1259.4... = 1260 cm^2 [1]

ACKNOWLEDGEMENTS

Every effort has been made to trace copyright holders and obtain their permission for the use of copyright material. The author and publisher will gladly receive information enabling them to rectify any error or omission in subsequent editions. All facts are correct at time of going to press.

Published by Collins
An imprint of HarperCollins*Publishers*
1 London Bridge Street
London SE1 9GF

HarperCollins*Publishers*
Macken House
39/40 Mayor Street Upper
Dublin 1
D01 C9W8
Ireland

© HarperCollins*Publishers* Limited 2021

ISBN 9780008326692

First published 2015

This edition published 2021

10 9 8 7 6 5 4

British Library Cataloguing in Publication Data.

A CIP record of this book is available from the British Library.

Publishers: Katie Sergeant and Clare Souza
Project Leader: Richard Toms
Project Management: Rebecca Skinner
Author: Trevor Senior
Cover Design: Kevin Robbins
Inside Concept Design: Sarah Duxbury and Paul Oates
Text Design and Layout: Jouve India Private Limited
Production: Karen Nulty
Printed in the United Kingdom

MIX
Paper | Supporting responsible forestry
FSC
www.fsc.org   FSC™ C007454

This book contains FSC™ certified paper and other controlled sources to ensure responsible forest management.

For more information visit: www.harpercollins.co.uk/green